AF305143

SISSA Springer Series

Volume 7

The **SISSA Springer Series** publishes research monographs, contributed volumes, conference proceedings and lectures notes in English language resulting from workshops, conferences, courses, schools, seminars, and research activities carried out by SISSA: https://www.sissa.it/.

The books in the series will discuss recent results and analyze new trends focusing on the following areas: geometry, mathematical analysis, mathematical modelling, mathematical physics, numerical analysis and scientific computing, showing a fruitful collaboration of scientists with researchers from other fields.

The series is aimed at providing useful reference material to students, academic and researchers at an international level.

Sandro Zagatti

Functional Analysis

Foreword by Maria Teresa Rotolo, Antonio Lerario

Sandro Zagatti (Deceased)
Department of Mathematics
SISSA
Trieste, Italy

ISSN 2524-857X   ISSN 2524-8588 (electronic)
SISSA Springer Series
ISBN 978-3-032-24474-1   ISBN 978-3-032-24475-8 (eBook)
https://doi.org/10.1007/978-3-032-24475-8

This Springer imprint is published by the registered company Springer Nature Switzerland AG
The registered company address is: Gewerbestrasse 11, 6330 Cham, Switzerland

If disposing of this product, please recycle the paper.

*Sandro Zagatti has died before the
publication of this book.*

Foreword

These lecture notes collect the material from a course in *Functional Analysis* originally taught by our colleague, Sandro Zagatti, at SISSA. The notes were based on the handwritten material Sandro used throughout his career—a style that reflected his reserved and methodical nature. Following his untimely passing, the Mathematics Area at SISSA recognized the enduring value of his teaching and took the initiative to preserve it. The notes have now been transcribed and typeset in LaTeX with great care to maintain the clarity and precision that defined Sandro's lectures.

Although Sandro was a quiet and private presence within the Mathematics Area, his dedication to teaching was well recognized. He left a deep and lasting impression on his students through lectures that were both rigorous and elegant.

We each had the privilege of attending this course—almost 20 years apart—and yet our experiences were remarkably similar: a clear and structured exposition, a focus on essential examples, and a rare ability to make subtle concepts accessible. In preparing these notes, we have preserved the structure of the course as we experienced it. Particular care was taken to present all proofs and examples in full detail. Sandro's approach—characterized by attention to detail and enriched by a wealth of simple, illuminating examples—continues to support students in developing independent reasoning and in mastering both the theoretical and practical aspects of the subject.

In this spirit, the exercises at the end of the notes are drawn from past written exams that Sandro himself provided to students for practice. Some of these exercises are echoed in the main text, either partially or fully worked out as part of the exposition. The exercises are presented in no particular order, as many require synthesizing concepts from multiple chapters, and we felt that forcing a thematic organization would have been artificial.

The bibliography, kept intentionally concise, reflects the references suggested by Sandro during the course. The primary reference remains *Functional Analysis* by H. Brezis, which we also recommend for further exercises and theoretical complements. The book by Checcucci, Tognoli, and Vesentini is suggested as a useful companion for any necessary background in topology, measure theory, and related areas not treated in detail in these notes.

Our hope is that these notes will remain as instructive and elegant for future readers as they were for us—as enduring in their content as in their style.

Trieste, Italy   Maria Teresa Rotolo
April 2025   Antonio Lerario

Acknowledgments Antonio Lerario and Maria Teresa Rotolo wish to thank (in alphabetical order) Giuseppe Brusca, Davide Donati, Antonio Milosh Radakovic, and Matteo Zanardini, for their careful reading of the manuscript and for their valuable comments and indications to improve it.

Competing Interests The authors have no competing interests to declare that are relevant to the content of this manuscript.

Contents

Symbols

$d_X(a, b)$	Distance in the space X between the points a and b, see Definition 2.1
$B(x, r)$	Ball centered at x of radius r, see 2.1
B_1^X	Unitary ball in the space X centered at zero, i.e., $B_1^X = B(0, 1)$, see Example 2.26
τ_X	Topology of the distance on the space X, see 2.2
A^c	For A subset of some set X, the complement of A in X, see Definition 2.3
$\bar{A}$	The closure of the set A, defined in 2.3
$\mathbb{K}$	Denotes an algebraic field that can be indifferently $\mathbb{R}$ or $\mathbb{C}$, see Section 2.2
$\| \cdot \|_X$	A norm over the vector space X, see Definition 2.9
$\mathrm{lsp}(E)$	The linear span of the set E, see Definition 2.14
$\mathrm{lsp}_{\mathbb{Q}}(E)$	The rational-coefficients linear span of the set E, see Definition 2.14
$[a, b]$	The segment with extremes a and b, see example 2.15
$\mathcal{P}^k(I)$	Space of polynomials of degree $\leq k$ and coefficients in I, see Example 2.15
$\mathcal{P}(I)$	Space of polynomials of any degree and coefficients in I, see Example 2.16
$\mathcal{C}(I)$	Space of continuous, real valued functions over the interval I, see Example 2.18
$\mathcal{C}(X)$	Space of continuous, complex valued functions over the compact space X, see 4.2
$\mathcal{C}(X; \mathbb{R})$	Space of continuous, real valued functions over the compact space X, see 4.2
$\mathcal{L}(X, Y)$	The space of linear bounded operators from X to Y, see Definition 3.5
$\mathcal{L}(X)$	$\mathcal{L}(X) = \mathcal{L}(X, X)$ first appears in Section 3.4
X'	Dual of the space X, i.e., $X' = \mathcal{L}(X, \mathbb{K})$, see Definition 3.7
$T|_X$	The restriction of the map T to the subset X, see Theorem 3.8

$\langle f, x\rangle_{X',X}$	For $x \in X$ and $f \in X'$, is the evaluation of the functional f in the point x, see 3.6
$a \cdot b$	Scalar product between the vectors a and b of $\mathbb{R}^n$, see Equation 3.7
$\mathrm{dom}(e)$	Notation for the domain of the function e, first appears in 3.12
$\mathrm{dist}(\cdot, A)$	Distance from the set A, defined in 3.19
p	Minkowski functional or gauge, defined in 3.26
X''	Bidual space of X, see Definition 3.22
$f \circ g$	Composition of the function f with the function g, first used in Lemma 3.52
$\mathcal{K}(X, Y)$	Set of compact bounded operators from X to Y, first in Lemma 3.53
Id	Denotes the identity function: $Id(x) = x$. First appears in Corollary 3.38
$\|(., .)\|_{X \times Y}$	Product norm on $X \times Y$, see Definition 3.39
$\Gamma(T)$	The graph of the operator T, see Definition 3.40
$\overline{Co(A)}$	The closed convex hull of the set A, see Definition 3.46
$u \vee v$	Maximum function, see 4.9
$u \wedge v$	Minimum function, see 4.9
$u^{\pm}$	Positive and negative part of u, see 4.9
$(\cdot, \cdot)$	Notation for a scalar product, first in Definition 5.1
$x \perp y$	Indicates that the vectors x and y are orthogonal, see Section 5.1
δ_{ij}	Indicator function: $\delta_{ij} = 1$ if $i = j$ and zero otherwise. First in the proof of Lemma 5.3
$\mathcal{H}$	Notation for a Hilbert space. See Section 5
$P_C x$	Projection of x over C, see Proposition 5.11
$S^{\perp}$	The orthogonal complement of S, see Definition 5.13
$\oplus$	Direct sum, first in Proposition 5.15
T^*	Adjoint of the operator T, see Section 5.5
ℓ^p	Space of p-summable sequences, see Definition 6.1
$\mathcal{C}_b(\Omega)$	Set of continuous and bounded functions on Ω, see 7.1
$\mathrm{supp}(u)$	Support of the function u, see Definition 7.1
$\mathcal{C}_c(\Omega)$	Set of continuous functions with bounded support in Ω, see Definition 7.2
$\mathcal{C}_0(\Omega)$	Closure of $\mathcal{C}_c(\Omega)$ in $\mathcal{C}_b(\Omega)$, see Definition 7.4
L^p	L^p space, see Definition 7.6
p'	Notation for the index conjugated to p, see Lemma 7.9
χ_Ω	Characteristic function of the set Ω. First used in the proof of Proposition 7.10
$\check{f}, \tau_a(f)$	See Section 7.2
$f * g$	Convolution between f and g, defined in 7.9
L^p_c	The set of L^p functions with compact support, first in Section 7.2
L^p_{loc}	Measurable functions, locally L^p on each compact set. First in 7.2
$\mathcal{C}^k_c(\Omega), \mathcal{C}^\infty_c(\Omega)$	Defined just before Proposition 7.19
$\nabla(.)$	Gradient of a function, used first in Proposition 7.19

τ_s	The topology induced by the norm in a Banach space, see Section 8.1
$\sigma(X, X')$	Weak topology on X, see Definition 8.3
$x_n \rightharpoonup x$	Weak convergence of the sequence x_n to the point x, see Definition 8.3
$\sigma(X', X)$	Weak* topology on the space X, see Definition 8.13
$x_n \overset{*}{\rightharpoonup} x$	Weak* convergence of the sequence x_n to the point x, see Definition 8.13
$\ker(T)$	Kernel of the operator T, first appears in the proof of Theorem 5.23
$\mathrm{Ran}(T)$	Image or range of the operator T, first appears in Definition 9.1
$\sigma(T)$	Spectrum of the operator T, see Definition 9.4
$\rho(T)$	Resolvent set of T, see Definition 9.4
$\sigma_p(T)$	Point spectrum of T, see Remark 9.5
E_λ	Eigenspace associated with the eigenvalue λ, see Remark 9.5
$r(T)$	Spectral radius of the operator T, see 9.13
$R(\lambda, T)$	Resolvent operator of T at λ, see Definition 9.7

Chapter 1
Introduction

The goal of this course is to introduce the basic tools of functional analysis. As the name suggests, the central objects of study of this field are functions, which, differently from classical analysis, are not studied in isolation but within the context of spaces that impose structure and allow for the analysis of their properties. Such spaces are called *function spaces*, and are essentially sets of functions endowed with additional structures such as norms, topologies, and other properties that facilitate operations and transformations.

Before delving into the technical aspects, let us motivate the study of functional analysis with a typical example in which functional spaces naturally appear. Consider the Cauchy problem

$$\begin{cases} \dot{y}(t) = f(t, y(t)), \\ y(t_0) = y_0, \end{cases} \tag{1.1}$$

where $f : \mathbb{R}^2 \to \mathbb{R}$ is a continuous and differentiable function. This differential equation can be transformed in the associated integral equation, equivalent to the Cauchy problem, which reads:

$$y(t) = y_0 + \int_{t_0}^{t} f(\tau, y(\tau))\mathrm{d}\tau.$$

In order to find solutions of the integral equation we fix $h \in \mathbb{R}$ and introduce the set

$$X := \{u : [t_0 - h, t_0 + h] \to \mathbb{R}, \quad u \text{ continuous}\}$$

and the map

$$\Phi : X \to X, \qquad \Phi(y) := y_0 + \int_{t_0}^{t} f(\tau, y(\tau))\mathrm{d}\tau.$$

A function $\overline{y} : [t_0 - h, t_0 + h] \to \mathbb{R}$ such that $\Phi(\overline{y}) = \overline{y}$, (i.e., a fixed point of Φ) is a solution of the Cauchy problem and viceversa. It is possible to prove that, if $|h|$ is sufficiently small, then $\Phi : X \to X$ has a fixed point. Thus, the strategy to solve the Cauchy problem (1.1) is to transform it into the equation $\Phi(y) = y$, which is a *functional equation*: the variable y is a function ranging in the *function space X*.

The main reason why function spaces are of particular interest is that the most used ones are infinite-dimensional spaces (as the one in the example above) and it turns out that passing from a finite-dimensional setting to an infinite dimensional one crucially changes the tools that one has at his disposal. For instance, in finite-dimensional spaces, all norms are equivalent, meaning that they define the same topology. However, as we will see many times in these lecture notes, in infinite-dimensional spaces, this is not the case, and different norms can lead to different topological properties. Another huge difference is that, in finite-dimensional spaces, a subset is compact if and only if it is closed and bounded (this result is typically known as the Heine-Borel Theorem). This characterization does not hold in infinite-dimensional spaces: for example, the unitary ball centered at the origin is not compact anymore. Thus, infinite-dimensional function spaces often require the introduction of multiple topologies, some of which may not be induced by a norm.

The ability to consider different, non-equivalent norms and distinct topologies in turn creates an interplay between open and compact sets. A more intricate topology leads to a smaller number of compact sets and conversely, a simpler topology results in a larger set of compact sets. This flexibility allows one to select the most appropriate topological framework for each specific problem, thus facilitating the understanding of its relationship with other frameworks. The primary objective of these lecture notes is to formalize this concept systematically. We will begin by exploring classical functional spaces, such as continuous and differentiable functions, which form the foundation of our understanding. From there, we will gradually expand our scope, introducing a variety of functional setups and topological structures. Each section builds on the previous ones, creating a comprehensive toolkit to tackle a wide range of mathematical problems.

Chapter 2
Basic Notions

We start with the basic definitions and results: we introduce the theory of metric and normed spaces, together with the first example of function space. We also use this section to give some definitions and set some notation that we will use in the rest of the notes.

2.1 Metric Spaces

Definition 2.1 (*Metric space*) Let X be a set and let $d_X : X \times X \to \mathbb{R}$ a map with the following properties:

- $d_X(x, y) \geq 0, \quad \forall x, y \in X, \quad$ and $\quad d_x(x, y) = 0 \iff x = y$;
- $d_X(x, y) = d_X(y, x), \quad \forall x, y \in X$;
- $d_X(x, y) \leq d_X(x, z) + d_X(z, y), \quad \forall x, y, z \in X$.

The map d_X is said to be a *distance* or a *metric* and the pair (X, d_X) is called a *metric space*.

Let $Y \subset X$ and set $d_Y(x, y) := d_X(x, y), \ \forall x, y \in Y$. Then the pair (Y, d_Y) is also a metric space and d_Y is called distance in Y induced by d_X.

When it is clear from the context we will just denote by d the distance d_X on the metric space X.

Definition 2.2 (*Complete metric space*) Let (X, d) be a metric space, $(x_n)_n$ a sequence in X and $x \in X$. We say that

1. $(x_n)_n$ *converges* to x and write $x_n \to x$ if $d(x_n, x) \to 0$, as $n \to \infty$;
2. $(x_n)_n$ is a *Cauchy sequence* if for all positive ε there exists a natural number n_ε such that
$$d(x_n, x_m) < \varepsilon, \quad \forall n, m \geq n_\varepsilon,$$

© The Author(s), under exclusive license to Springer Nature Switzerland AG 2026
S. Zagatti, *Functional Analysis*, SISSA Springer Series 7,
https://doi.org/10.1007/978-3-032-24475-8_2

in other words, $d(x_n, x_m)$ tends to zero as n, m tend to infinity.

We say that the space (X, d) is *complete* if any Cauchy sequence in X converges to some element of X.

Definition 2.3 (*Open sets and topology on X*) Let (X, d) be a metric space, $x_0 \in X$, $r > 0$. We set

$$B(x_0, r) := \{y \in X : d(y, x_0) < r\} \qquad (2.1)$$

and call $B(x_0, r)$ the *open ball* in X of center x_0 and radius r. A nonempty subset A of X is said to be *open* in X if for all y in A there exist a positive r such that the ball $B(y, r)$ centered in y and of radius r is contained in A. The family

$$\tau_X := \{A \subseteq X : A \text{ is open in } X\} \cup \{\emptyset\} \qquad (2.2)$$

is called topology induced by the distance d or tolopology of the distance in (X, d). If $C \subseteq X$ is such that $C^c := X \setminus C \in \tau_X$, then we say that C is *closed*. In particular every open ball is open and $\emptyset$ and X are both open and closed.

Definition 2.4 Let (X, d) be a metric space and $A \subseteq X$. We set

$$\overline{A} = \bigcap_{\mathcal{E}} C, \quad \mathcal{E} := \{C \subseteq X, C \text{ closed}, \ A \subseteq C\} \qquad (2.3)$$

and call $\overline{A}$ *the closure* of A. We remark that $\overline{A}$ is closed.

Definition 2.5 Let (M, d_M) and (N, d_N) be metric spaces and let $f : M \to N$. We say that f is *continuous* if for every open subset $A \subseteq N$ its preimage $f^{-1}(A)$ is open in M.

Definition 2.6 (*Density and separability*) Let (M, d) be a metric space.

1. A subset $A \subseteq M$ is *dense* in M if for every element y of M and for every positive r, there exists an element x of A which is contained in the ball $B(y, r)$. Equivalently a subset A of M is dense if for any y in M there exists a sequence $(x_n)_n$ in A such that $x_n \to y$. In particular if A is dense, then $\overline{A} = M$.
2. The space M is *separable* if there exists a subset A of M which is both countable and dense in M.

Theorem 2.7 *Let (M, d) be a (non-complete) metric space. There exists a complete metric space $(\tilde{M}, \tilde{d})$ and a map $h : M \to \tilde{M}$ such that*

1. $\tilde{d}(h(x), h(y)) = d(x, y), \quad \forall x, y \in M$;
2. $h(M)$ is dense in $\tilde{M}$.

$(\tilde{M}, \tilde{d})$ *is called the completion of* (M, d).

Proof We just sketch the proof, since many other similar completion theorems will be stated in the following sections, and the proofs always follow the same idea.

Let

$$\mathcal{M} := \{(x_n)_n : \text{Cauchy sequence in } M\}$$

and given $(x_n)_n, (y_n)_n$ in $\mathcal{M}$ define the equivalence relation $(x_n)_n \sim (y_n)_n$ if $d(x_n, y_n) \to 0$. Then set $\tilde{M} := \mathcal{M}/\sim$. Given $\tilde{M} \ni \xi = [(x_n)_n]$, $\tilde{M} \ni \eta = [(y_n)_n]$, we set $\tilde{d}(\xi, \eta) := \lim_{n\to\infty} d(x_n, y_n)$. Given $x \in M$, we consider the constant sequence $x_n = x$ for all $n \in \mathbb{N}$ and set $h(x) := [(x_n)_n] \in \tilde{M}$. Finally, one can check that $\tilde{M}, \tilde{d}$ and $\tilde{h}$ satisfy the desired properties. $\qquad\square$

Example 2.8 Consider the metric space $\mathbb{Q}$ of rational numbers, equipped with the usual distance given by the absolute value of the difference between two rational numbers: $d_{\mathbb{Q}}(p, q) := |q - p|$, for all $p, q \in \mathbb{Q}$. The space $(\mathbb{Q}, d_{\mathbb{Q}})$ is not complete, and its completion is $\mathbb{R}$, which is a metric space if endowed with the same distance $d_{\mathbb{R}}(r, s) := |s - r|$, for all $r, s \in \mathbb{R}$.

2.2 Normed Spaces

We now introduce *normed spaces*. From now on, we will write $\mathbb{K}$ to denote a field that can indifferently be $\mathbb{K} = \mathbb{R}$ or $\mathbb{K} = \mathbb{C}$.

Definition 2.9 (*Normed space*) Let X be a vector space over $\mathbb{K}$ and let $\|\cdot\| : X \to \mathbb{R}$ be a map with the following properties:

1. $\|x\| \geq 0 \,\forall x \in X, \quad \|x\| = 0 \iff x = 0$;
2. $\|\lambda x\| = |\lambda| \,\|x\|, \forall x \in X, \lambda \in \mathbb{K}$;
3. $\|x + y\| \leq \|x\| + \|y\|, \quad \forall x, y \in X$.

The function $\|\cdot\|$ is a *norm* over X and the pair $(X, \|\cdot\|)$ is a *normed space*.

Remark 2.10 Any normed space is a metric space: setting indeed $d(x, y) := \|x - y\|$, one can easily check that d satisfies the properties of Definition 2.1. Hence for normed spaces we can express open balls in (2.1) as $B(x_0, r) = \{y \in X : \|x_0 - y\| < r\}$. Thus, being in a metric setting, it makes sense to remark that a norm is a continuous function. Indeed:

$$\|x\| \leq \|x - y + y\| \leq \|x - y\| + \|y\| \implies \Big|\|x\| - \|y\|\Big| \leq \|x - y\| = d(x, y), \quad \forall x, y \in X.$$

Thanks to Remark 2.10 we can now easily rephrase Theorem 2.7 for normed spaces:

Corollary 2.11 *Let $(X, \|\cdot\|)$ be a non-complete normed space. There exists a complete normed space $(\tilde{X}, \|\cdot\|_{\tilde{X}})$ and a map $h : X \to \tilde{X}$ such that*

1. *$\|h(x)\|_{\tilde{X}} = \|x\|, \quad \forall x \in X$;*
2. *$h(X)$ is dense in $\tilde{X}$.*

Definition 2.12 (*Equivalent norms*) Two different norms $\|\cdot\|_1$ and $\|\cdot\|_2$ on a vector space X are *equivalent* if there exist two positive constant C_1 and C_2 such that

$$C_1 \|x\|_1 \le \|x\|_2 \le C_2 \|x\|_1 .$$

Example 2.13 If $X = \mathbb{R}$, we define $\|x\| := |x|$. Let now $X := \mathbb{R}^n$, $x = (x_1, \ldots, x_n) \in X$. We can endow $\mathbb{R}^n$ with different norms, such as

$$\|x\|_e := \left(\sum_{i=1}^{n} x_i^2 \right)^{\frac{1}{2}}, \quad \|x\|_1 := \sum_{i=1}^{n} |x_i|, \quad \|x\|_\infty = \max_i |x_i|.$$

It is possible to prove that on any finite dimensional space two different norms are always equivalent, thus, all the norms defined in this example are equivalent in $\mathbb{R}^n$.

Definition 2.14 Let X be a vector space over $\mathbb{K}$. Given $E \subseteq X$ we define the *linear span* of E as

$$\mathrm{lsp}(E) := \{z = \sum_{i=1}^{N_z} \lambda_i y_i, \ y_i \in E, \ \lambda_i \in \mathbb{K}, \ N_z \in \mathbb{N}\}.$$

If there exists $E \subseteq X$, finite set, such that $X = \mathrm{lsp}(E)$, we say that X is *finite dimensional*. We call *dimension* of X the cardinality of a basis of E. If X is not finite dimensional we say that it is *infinite dimensional*. We shall also use the following notation:

$$\mathrm{lsp}_{\mathbb{Q}}(E) := \{z = \sum_{i=1}^{N_z} \lambda_i y_i, \ y_i \in E, \ \lambda_i \in \mathbb{Q}(\text{ or } \mathbb{Q} + i\mathbb{Q})\}.$$

2.3　A First Example

We now define the finite dimensional function space $\mathcal{P}^k(I)$ of polynomials over a real interval I with real coefficients and degree smaller or equal than a fixed natural number k. We endow this space with two different norms (which are equivalent since the space is finite dimensional) and then we show that it is a complete space.

Example 2.15 Let $I = [\alpha, \beta] \subset \mathbb{R}$ be a closed and bounded interval. We set

$$\mathcal{P}^k(I) := \{p : I \to \mathbb{R}, \ p(t) = \sum_{j=0}^{k} a_j t^j, a_j \in \mathbb{R}\}. \tag{2.4}$$

$\mathcal{P}^k(I)$ is a linear space of dimension $k + 1$. Setting $E = \{e_j(t) := t^j, \ j = 0, \ldots, k\}$ we have $\mathcal{P}^k(I) = \mathrm{lsp}(E)$, according to Definition 2.14. It is easy to see that the following map is a norm over $\mathcal{P}^k(I)$:

$$\|p\|_1 := \left(\sum_{j=0}^{k} a_j^2\right)^{\frac{1}{2}}.$$

(2.5)

Let us also define

$$\|p\|_\infty := \max_{t\in I} |p(t)|.$$

(2.6)

We show that it is a norm over $\mathcal{P}^k(I)$:

1. $\|p\|_\infty = \max_{t\in I} |p(t)| \geq 0, \quad \forall p \in \mathcal{P}^k(I),$ and $\max_{t\in I} |p(t)| = 0 \iff p \equiv 0$;
2. $\|\lambda p\|_\infty = \max_{t\in I} |\lambda p(t)| = |\lambda| \max_{t\in I} |p(t)| = |\lambda| \|p\|_\infty$;
3.

$$\|p+q\|_\infty = \max_{t\in I} |p(t)+q(t)| \leq \max_{t\in I}(|p(t)|+|q(t)|) \leq \max_{t\in I} |p(t)| + \max_{t\in I} |q(t)|$$
$$= \|p\|_\infty + \|q\|_\infty.$$

(2.7)

Hence $(\mathcal{P}^k(I), \|\cdot\|_\infty)$ is a normed space. Since $\mathcal{P}^k(I)$ is finite dimensional, the two norms $\|\cdot\|_1$ and $\|\cdot\|_\infty$ are equivalent.

We now conclude the example proving that $\mathcal{P}^k(I)$ is complete: we have to show that every Cauchy sequence converges to an element of $\mathcal{P}^k(I)$: let $(p_n)_n$ be a Cauchy sequence in $\mathcal{P}^k(I)$, i.e.,

$$\|p_n - p_m\|_1 \to 0, \text{ as } n, m \to \infty, \text{ or equivalently } \|p_n - p_m\|_\infty \to 0, \text{ as } n, m \to \infty. \quad (2.8)$$

Denoting by $a^n \in \mathbb{R}^{k+1}$ the vector of coefficients of p_n, Eq. (2.8) reads

$$\left\|a^n - a^m\right\|_1 \to 0, \quad \text{or} \quad \left\|a^n - a^m\right\|_\infty \to 0, \quad \text{as } n, m \to \infty.$$

Since $\mathbb{R}^{k+1}$ is a complete space, there exists $a = (a_0, \ldots, a_k) \in \mathbb{R}^{k+1}$ such that $a^n \to a$ in $\mathbb{R}^{k+1}$. Set $p(t) := \sum_{j=0}^{k} a_j t^j$. We immediately see that $\|p_n - p\|_1 \to 0$ (or equivalently that $\|p_n - p\|_\infty \to 0$), as $n \to \infty$. Hence $\mathcal{P}^k(I)$ is complete.

We remark that in this example we strongly used that $\mathcal{P}^k(I)$ is isomorphic to $\mathbb{R}^{k+1}$. It is a general fact indeed that all finite dimensional vector spaces are metrically equivalent to $\mathbb{R}^m$ for some natural m. In particular, for every finite dimensional vector space M of dimension m with distance d_M, there exists a linear bijection between $\mathbb{R}^m$ and M that preserves the distance functions. We now turn to infinite dimensional spaces: we construct the first infinite dimensional function space starting from Example 2.15.

Example 2.16 Consider the following linear space over $\mathbb{R}$:

$$\mathcal{P}(I) := \bigcup_{k=0}^{\infty} \mathcal{P}^k(I), \tag{2.9}$$

with $\mathcal{P}^k(I)$ defined in (2.4). Clearly $\mathcal{P}(I)$ is infinite dimensional since it contains the spaces $\mathcal{P}^k(I)$, for all k. We define $\|p\|_{\mathcal{P}(I)} := \|p\|_{\infty}$, (see (2.6)).

The same computations done in Example 2.15 show that $(\mathcal{P}(I), \|\cdot\|_{\infty})$ is a normed space. We now show that it is *not complete*. Consider the sequence $(p_n)_n$ given by $p_n(t) := \sum_{j=0}^{n} \frac{t^j}{j!}$. Clearly $p_n \in \mathcal{P}(I)$ for all $n \in \mathbb{N}$ (here we use the convention that $0! = 1$). Computing

$$|p_{n+k}(t) - p_n(t)| = \Big| \sum_{j=n+1}^{n+k} \frac{t^j}{j!} \Big| \leq \sum_{j=n+1}^{n+k} \frac{|t^j|}{j!} \leq \sum_{j=n+1}^{\infty} \frac{|t^j|}{j!}$$

we see that

$$\|p_{n+k} - p_n\|_{\infty} \leq \sum_{j=n+1}^{\infty} \frac{c^j}{j!} \to 0, \quad \text{as } n \to \infty,$$

with $c := \max\{|a|, |b|\}$ (recall that a and b are the extremes of the interval I). This computation shows that $(p_n)_n$ is a Cauchy sequence in $\mathcal{P}(I)$, but since $p_n(t)$ converges to the exponential function $f(t) := e^t$, which is not an element of $\mathcal{P}(I)$, the space $\mathcal{P}(I)$ is not complete. Thus, we now ask what is the completion of the space $\mathcal{P}(I)$, i.e., we look for the minimal complete normed space containing $\mathcal{P}(I)$.

Definition 2.17 (*Banach space*) A complete normed space is called *Banach space*.

Example 2.18 (*The Banach space of continuous functions on an interval*) Let $I = [a, b] \subset \mathbb{R}$ be a closed and bounded interval. We set

$$\mathcal{C}(I) = \{u : I \to \mathbb{R}, \ u \text{ continuous}\}.$$

Clearly $\mathcal{C}(I)$ is a linear space and it is infinite dimensional since $\mathcal{P}(I) \subset \mathcal{C}(I)$. We set, for any u in $\mathcal{C}(I)$, the norm of u as $\|u\|_{\mathcal{C}(I)} := \|u\|_{\infty}$ (see (2.6)). Using the same computations of Example 2.15 one can prove that $(\mathcal{C}(I), \|\cdot\|_{\infty})$ is a normed space. We now prove that it is also complete, constructing the first example of an infinite dimensional Banach space. Let $(u_n)_n$ be a Cauchy sequence in $\mathcal{C}(I)$, we claim that there exists $u \in \mathcal{C}(I)$ such that

$$\|u_n - u\|_{\mathcal{C}(I)} \to 0, \quad \text{as} \quad n \to +\infty. \tag{2.10}$$

For every fixed $x \in I$ we have that

$$|u_n(x) - u_m(x)| \leq \max_{t \in I} |u_n(t) - u_m(t)| \to 0, \quad \text{as} \quad n, m \to \infty,$$

thus, $(u_n(x))_n$ is a Cauchy sequence in $\mathbb{R}$, and from completeness of $\mathbb{R}$ there exists $u(x) \in \mathbb{R}$ such that $u_n(x)$ converges to $u(x)$, as n tends to infinity. Defining the function u as the one that associates to each x in the interval I the value $u(x)$, we show that condition (2.10) is satisfied:

$$|u_n(x) - u(x)| = \lim_{m \to \infty} |u_n(x) - u_m(x)| \le \lim_{m \to \infty} \|u_n - u_m\|_\infty ,$$

thus, given any positive ε there exists a natural number n_ε such that $|u_n(x) - u(x)| \le \varepsilon$, for all $n \ge n_\varepsilon$. Since n_ε does not depend on x we obtain (2.10) and we say that u_n *converges uniformly* to u on I.

Finally we show that u is continuous: take any $t, s \in I$ and write

$$\begin{aligned}
u(t) - u(s) = & u(t) - u_n(t) \\
& + u_n(t) - u_n(s) \\
& + u_n(s) - u(s).
\end{aligned} \tag{2.11}$$

This implies that

$$|u(t) - u(s)| \le \|u - u_n\|_\infty + |u_n(t) - u_n(s)| + \|u - u_n\|_\infty .$$

Fix now ε arbitrarily small and choose $n \in \mathbb{N}$ large enough so that $\|u - u_n\|_\infty < \varepsilon$, which is possible from (2.10). Since u_n is uniformly continuous, there exists $\delta > 0$ such that $|u_n(t) - u_n(s)| \le \varepsilon$, for all t, s such that $|t - s| \le \delta$. Hence $|u(t) - u(s)| < 3\varepsilon$ and thus, $u \in \mathcal{C}(I)$.

We conclude this example showing that we can introduce another norm over $\mathcal{C}(I)$, which is not equivalent to $\|\cdot\|_\infty$ and with respect to which $\mathcal{C}(I)$ is not complete: define

$$\|u\|_1 := \int_I |u(t)|\mathrm{d}t. \tag{2.12}$$

It is easy to check that (2.12) defines a norm over $\mathcal{C}(I)$. We now show that $\|\cdot\|_\infty$ and $\|\cdot\|_1$ are not equivalent: suppose for simplicity that $I = [0, 1]$. For $u \in \mathcal{C}(I)$ we have:

$$\|u\|_1 = \int_I |u(t)|\mathrm{d}t \le \int_I \max_I |u|\mathrm{d}t = \|u\|_\infty .$$

The opposite inequality is instead false in general: consider the sequence $u_n(t) := t^n$, for $t \in [0, 1]$. Clearly the sequence belongs to $\mathcal{C}(I)$, since all the u_n are continuous functions over $[0, 1]$. Moreover, $\|u_n\|_\infty = 1$ for all n. But

$$\|u_n\|_1 = \int_0^1 t^n \mathrm{d}t = \frac{1}{n+1} \to 0, \quad \text{as} \quad n \to \infty.$$

Table 2.1 Previous examples in a glimpse

Normed space	Dimension	Completeness
$(\mathcal{P}^k(I),\ \text{any}\ )$	k	YES
$(\mathcal{P}(I), \|\cdot\|_\infty)$	∞	NO
$(\mathcal{C}(I), \|\cdot\|_\infty)$	∞	YES
$(\mathcal{C}(I), \|\cdot\|_1)$	∞	NO

Finally we show that $(\mathcal{C}(I), \|\cdot\|_1)$ is not complete. It is sufficient to find a single Cauchy sequence $(u_n)_n$ of continuous functions which does not converge to an element of $\mathcal{C}(I)$ in norm $\|\cdot\|_1$. Take again $I = [0, 1]$ and $u_n(t) = t^n$ (Table 2.1).

$$\|u_{n+m} - u_n\|_1 = \int_0^1 |t^{n+m} - t^n| \mathrm{d}t \leq \int_0^1 t^n \mathrm{d}t = \frac{1}{n+1} \to 0, \quad \text{as } n \to +\infty.$$

Hence $(u_n)_n$ is a a Cauchy sequence, but

$$\|u_n - u\|_1 \to 0, \quad \text{as } n \to \infty \quad \text{with} \quad u(t) := \begin{cases} 0 & t \in [0, 1[\\ 1 & t = 1 \end{cases}$$

and $u \notin \mathcal{C}(I)$.

2.4 Compactness

Definition 2.19 (*Compact space*) Let (X, τ_X) be a topological space (where τ_X is defined as in (2.2)). A family $\mathcal{U} \subset \tau_X$ is called an open covering of X if

$$X = \bigcup_{U \in \mathcal{U}} U.$$

The space X is said to be *compact* if any open covering of X admits a finite sub-covering, i.e., if for every open covering $\mathcal{U} \subset \tau_X$ there exists a finite subfamily $\{U_1, \ldots, U_n\} \subseteq \mathcal{U}$ such that $X \subseteq \bigcup_{i=1}^n U_i$.

Remark 2.20 For any subset $Y \subseteq X$ of a topological space (X, τ_X), Y is compact if Definition 2.19 holds for the topological space (Y, τ_Y) where the induced topology τ_Y on Y is defined as $\tau_Y := \{Y \cap A, A \in \tau_X\}$. With this notation we also say that $Y \subset X$ is *relatively compact* if $\overline{Y}$ (see Definition (2.3)) is compact.

Definition 2.21 We say that:

1. X is *sequentially compact* if for any sequence $(x_n)_n \subset X$ there exists a subsequence $(x_{n_k})_k$ converging to some $x \in X$;

2. $Y \subseteq X$ is *relatively sequentially compact* in X if $\overline{Y}$ is sequentially compact.

The following is a classical result on metric spaces, thus, we do not prove it in these notes and we refer to [1] for a proof.

Proposition 2.22 *Let (X, d) be a metric space and let $Y \subseteq X$.*

(i) Y is compact if and only if Y is sequentially compact;
(ii) Y is relatively compact if and only if Y is relatively sequentially compact.

Definition 2.23 Let (X, d) be a metric space and let $Y \subseteq X$. Y is *precompact* (or *totally bounded*) if

$$\forall \varepsilon > 0 \; \exists \, N_\varepsilon \in \mathbb{N} \text{ and } y_1, \ldots, y_{N_\varepsilon} \in X \text{ such that } Y \subseteq \bigcup_{i=1}^{N_\varepsilon} B(y_i, \varepsilon).$$

Proposition 2.24 *Let (X, d) be a metric space and let $Y \subseteq X$.*

(i) If Y is relatively compact, then Y is precompact;
(ii) if (X, d) is complete and Y is precompact, then Y is also relatively compact.

Corollary 2.25 *Let (X, d) be a Banach space and let $Y \subseteq X$. Then Y is precompact if and only if Y is relatively compact.*

It is a well known result that if X is a finite dimensional normed space, the closed unit ball

$$\overline{B}_1^X := \{x \in X : \|x\| \leq 1\} \tag{2.13}$$

is compact (Heine-Borel Theorem). Let us now show instead an example of infinite dimensional space in which the unit ball is not compact:

Example 2.26 Consider the space $\mathcal{C}(I)$ introduced in Example 2.18 with $I = [0, 1]$. Then $\overline{B}_1^{\mathcal{C}(I)} = \{u \in \mathcal{C}(I) : \|u\|_\infty \leq 1\}$. Consider the sequence $u_n(t) = t^n$, with $t \in [0, 1]$. Clearly $u_n \in \overline{B}_1^{\mathcal{C}(I)}$, $\forall n \in \mathbb{N}$. But

$$u_n(t) \to u(t) := \begin{cases} 0, & t \in [0, 1[\\ 1, & t = 1, \end{cases}$$

which does not belong to $\mathcal{C}(I)$. Hence $\overline{B}_1^{\mathcal{C}(I)}$ is not compact.

Motivated by this example we now state an important property of compactness in normed spaces.

Theorem 2.27 *Let $(X, \|\cdot\|)$ be a normed space, then $\overline{B}_1^X$ defined in (2.13) is precompact if and only if X is finite dimensional.*

Proof We only show that if $\overline{B}_1^X$ is precompact then X is finite dimensional, since the other implication is the statement of the Heine-Borel Theorem. To this aim, following Definition 2.23 of precompactness, denote by $\{y_1, \dots, y_N\} \subset X$ a finite set of points such that $\overline{B}_1^X \subseteq \bigcup_{i=1}^N B\left(y_i, \frac{1}{2}\right)$ and remark that

$$\overline{B}_1^X \subseteq \bigcup_{i=1}^N B\left(y_i, \frac{1}{2}\right) \subseteq \bigcup_{i=1}^N \left(y_i + \frac{1}{2}\overline{B}_1^X\right).$$

Thus, define the N-dimensional linear subspace $Y := \mathrm{lsp}\{y_1, \dots, y_N\}$, so that $\overline{B}_1^X \subseteq Y + \frac{1}{2}\overline{B}_1^X$. Iterating:

$$\overline{B}_1^X \subseteq Y + \frac{1}{2}\overline{B}_1^X = Y + \frac{1}{2}Y + \frac{1}{4}\overline{B}_1^X = Y + \frac{1}{4}\overline{B}_1^X \ \dots$$

Consequently:

$$\overline{B}_1^X \subseteq Y + \bigcap_{n\in\mathbb{N}} \frac{1}{2^n}\overline{B}_1^X = Y.$$

We now show that $X \subset Y$. Let $z \in X$, $z \neq 0$ and remark that $z = \|z\| \frac{z}{\|z\|} = \|z\|\,\zeta$, with $\zeta \in \overline{B}_1^X$. Hence $\zeta \in Y$, implying by linearity of Y that $z \in Y$, proving the inclusion $X \subseteq Y$ that concludes the proof. $\qquad\square$

Reference

1. Checcucci, V., Tognoli, A., Vesentini, E.: Lezioni di topologia generale. Feltrinelli, Collana di matematica (1971)

Chapter 3
Continuous Linear Operators

We now introduce maps between normed spaces.

Definition 3.1 Let $(X, \|\cdot\|_X)$ and $(Y, \|\cdot\|_Y)$ be normed spaces over the field $\mathbb{K}$. The map $T : X \to Y$ is *linear* if

$$T(\alpha x + \beta y) = \alpha T x + \beta T y, \quad \forall x, y \in X, \quad \forall \alpha, \beta \in \mathbb{K};$$

the map T is *bounded* if there exists $M \geq 0$ such that

$$\|T x\|_Y \leq M \|x\|_X, \quad \forall x \in X. \tag{3.1}$$

Example 3.2 (*Linear maps in finite dimension*) Let $X := \mathbb{R}^n$ and $Y := \mathbb{R}^m$ where n and m are arbitrary natural numbers. Then for every $T : \mathbb{R}^n \to \mathbb{R}^m$ there exists $A = (a_{i,j})_{i=1,\dots,n}^{j=1,\dots,m} \in \mathcal{M}^{m \times n}$, (the space of matrices of dimension $m \times n$) real matrix such that $Tx = Ax$. We have that, denoting by $\|\cdot\|_{\mathbb{R}^n} := \|\cdot\|_e$ the euclidean norm, (see Example 2.13)

$$\|T x\|_{\mathbb{R}^m} \leq \|A\| \, \|x\|_{\mathbb{R}^n}, \quad \text{where} \quad \|A\| := \left(\sum_{i,j} a_{i,j}^2 \right)^{\frac{1}{2}}.$$

Proposition 3.3 *Let $T : X \to Y$ be a linear operator. Then the following are equivalent:*

 (i) T is continuous on X;
(ii) T is continuos at a point $x_0 \in X$;
(iii) T is bounded.

Proof Clearly (i) implies (ii). We now show that (ii) implies (iii): from the very definition of continuity, for every $\varepsilon > 0$ there exists $\delta > 0$ such that

$$\|x - x_0\|_X \leq \delta \quad \Longrightarrow \quad \|T x - T x_0\|_Y \leq \varepsilon.$$

© The Author(s), under exclusive license to Springer Nature Switzerland AG 2026
S. Zagatti, *Functional Analysis*, SISSA Springer Series 7,
https://doi.org/10.1007/978-3-032-24475-8_3

Thus, for an arbitrary $z \in X$, $z \neq 0$, we define $x_z := x_0 + \frac{1}{\|z\|_X}\delta z$, obtaining $\|x_z - x_0\|_X \leq \delta$ and thus, $\|Tx_z - Tx_0\| \leq \varepsilon$. Since by linearity $Tx_z = Tx_0 + \frac{\delta}{\|z\|_X}Tz$, we have

$$\|Tx_z - Tx_0\|_Y = \left\| \frac{\delta}{\|z\|_X}Tz \right\|_Y = \frac{\delta}{\|z\|_X}\|Tz\|_Y .$$

Using this last equality:

$$\frac{\delta}{\|z\|_X}\|Tz\|_Y < \varepsilon \quad \Longrightarrow \quad \|Tz\|_Y \leq \frac{\varepsilon}{\delta}\|z\|_X ,$$

and since this holds for every $z \in X$ we conclude the proof of this implication setting $M := \frac{\varepsilon}{\delta}$, so that M satisfies (3.1).

Finally we prove that (iii) implies (i): for every $x, y \in X$

$$\|Tx - Ty\|_Y = \|T(x - y)\|_Y \leq M \|x - y\|_X ,$$

from linearity of T. Then T is continuous on X. $\qquad\qquad\qquad\qquad\square$

Definition 3.4 For a linear and bounded operator $T : X \to Y$ we define its operatorial norm as

$$
\begin{aligned}
\|T\| : &= \sup_{x \neq 0} \frac{\|Tx\|_Y}{\|x\|_X} \\
&= \sup_{\substack{\|x\|_X \leq 1 \\ x \neq 0}} \|Tx\|_Y \qquad\qquad\qquad (3.2) \\
&= \sup_{\|x\|_X = 1} \|Tx\|_Y \\
&= \inf\{M \geq 0 : \|Tx\|_Y \leq M \|x\|_X , \quad \forall x \in X\}.
\end{aligned}
$$

Exercise: prove all identities in (3.2).

Definition 3.5 We define the space $\mathcal{L}(X, Y) := \{T : X \to Y, \text{ linear and bounded}\}$ and we set $\|T\|_{\mathcal{L}(X,Y)} := \|T\|$ as in Definition 3.4.

Lemma 3.6 $(\mathcal{L}(X, Y), \|\cdot\|_{\mathcal{L}(X,Y)})$ *is a normed space over* $\mathbb{K}$.

Proof We immediately check that $\mathcal{L}(X, Y)$ is a linear space: indeed for any $T, S \in \mathcal{L}(X, Y)$ and for any $\alpha \in \mathbb{K}$

$$(T + S)x := Tx + Sx \in \mathcal{L}(X, Y), \quad (\alpha T)x := \alpha(Tx) \in \mathcal{L}(X, Y).$$

Now we show that $\|\cdot\|_{\mathcal{L}(X,Y)}$ is a norm: clearly if $\|T\| = 0$ then $\|Tx\|_Y = 0$, for all $x \in X$, thus, $T \equiv 0$; next

$$\|\lambda T\| \overset{3.2}{=} \sup_{\|x\|_X = 1} \|\lambda Tx\|_Y = |\lambda| \sup_{\|x\|_X = 1} \|Tx\|_Y = |\lambda| \|T\| .$$

Finally triangular inequality holds:

$$\|T + S\| = \sup_{\|x\|_X = 1} \|(T + S)x\|_Y = \sup_{\|x\|_X = 1} \|Tx + Sx\|_Y \le \sup_{\|x\|_X = 1} \|Tx\|_Y + \sup_{\|x\|_X = 1} \|Sx\|_Y = \|T\| + \|S\|,$$

$$(3.3)$$

concluding the proof. $\qquad\square$

Definition 3.7 (*Dual space*) Let $(X, \|\cdot\|_X)$ be a normed space over $\mathbb{K}$. We set $X' := \mathcal{L}(X, \mathbb{K})$ and call X' the dual space of X.

Theorem 3.8 (Linear extension) *Let $(X, \|\cdot\|_X)$ be a non-complete normed space and let $(Y, \|\cdot\|_Y)$ be a Banach space. Let $(\tilde{X}, \|\cdot\|_{\tilde{X}})$ be the completion of $(X, \|\cdot\|_X)$ defined by Corollary 2.11. Given $T \in \mathcal{L}(X, Y)$ there exists a unique $\tilde{T} \in \mathcal{L}(\tilde{X}, Y)$ such that $\tilde{T}|_X = T$. In addition*

$$\left\| \tilde{T} \right\|_{\mathcal{L}(\tilde{X},Y)} = \|T\|_{\mathcal{L}(X,Y)} . \qquad (3.4)$$

Remark 3.9 Writing $\tilde{T}|_X$ in Theorem 3.8, we mean $\tilde{T}|_{h(X)}$ where $h : X \to \tilde{X}$ is the isometric injection of X into $\tilde{X}$ given by Corollary 2.11.

Proof We sketch the proof referring to that of Theorem 2.7, where some definitions that we omit here are specified.

We define the map $\tilde{T}$ as follows: if $x \in X$ then $\tilde{T}x := Tx$. If $\tilde{x} \in \tilde{X} \setminus X$, we recall the definition of the space $\tilde{X}$ and consider the equivalence class $[(x_n)_n]$ defining $\tilde{x}$, where $(x_n)_n$ is a Cauchy sequence in X. Then the sequence $(Tx_n)_n$ is a Cauchy sequence in Y, since

$$\|Tx_n - Tx_m\|_Y = \|T(x_n - x_m)\|_Y \le \|T\| \, \|x_n - x_m\|_X .$$

Hence there exists $y \in Y$ such that $Tx_n \to y$ in Y, since Y is a complete space. Then we set $\tilde{T}\tilde{x} := y$.

We now check that $\tilde{T}$ is well-defined: take two sequences $(x_n)_n \sim (y_n)_n$ (i.e., $(y_n)_n \in [(x_n)_n]$), this implies that $\|x_n - y_n\|_X \to 0$, as $n \to \infty$. Then we have

$$\|Tx_n + Ty_n\| \le \|T\| \, \|x_n - y_n\| \to 0 \text{ as } n \to \infty,$$

hence $Ty_n \overset{n\to\infty}{\to} Ty$. Next we prove that $\tilde{T}$ is *linear*: let $\tilde{x} = [(x_n)_n]$ and $\tilde{y} = [(y_n)_n] \in \tilde{X}$, then $\tilde{x} + \tilde{y} = [(x_n + y_n)]$ and

$$\tilde{T}(\tilde{x} + \tilde{y}) = \lim_{n\to\infty} T(x_n + y_n) = \lim_{n\to\infty}(Tx_n + Ty_n) = \lim_{n\to\infty}(Tx_n) + \lim_{n\to\infty}(Ty_n) = \tilde{T}\tilde{x} + \tilde{T}\tilde{y}.$$

Analogously one shows that $\tilde{T}(\alpha\tilde{x}) = \alpha\tilde{T}\tilde{x}$, for all $x \in \tilde{X}$ and $\alpha \in \mathbb{K}$. Finally we show that $\tilde{T}$ is *continuous*: let $\tilde{x} \in \tilde{X}, \tilde{x} = [(x_n)_n]$. Then $\tilde{T}\tilde{x} = \lim_{n\to+\infty} Tx_n$. From the continuity of the norm

$$\left\|\tilde{T}\tilde{x}\right\|_Y = \lim_{n\to+\infty} \|Tx_n\|_Y \le \|T\| \lim_{n\to+\infty} \|x_n\| = \|T\|\,\|\tilde{x}\|_{\tilde{X}}.$$

Hence $\tilde{T} \in \mathcal{L}(\tilde{X}, Y)$ and $\left\|\tilde{T}\right\| \le \|T\|$. Since $\tilde{T}$ is an extension of T we also have $\|T\| \le \left\|\tilde{T}\right\|$, thus, proving that $\|T\| = \left\|\tilde{T}\right\|$.

We conclude the proof showing *uniqueness* of $\tilde{T}$: suppose that there exists $S \in \mathcal{L}(\tilde{X}, Y)$ satisfying the same properties as $\tilde{T}$. Then we have $\tilde{T} - S \in \mathcal{L}(\tilde{X}, Y)$ and $(\tilde{T} - S)|_X = 0$, which, from density of X in $\tilde{X}$ implies that $\tilde{T} - S \equiv 0$. $\square$

Theorem 3.10 *Let $(X, \|\cdot\|_X)$ be a normed space and let $(Y, \|\cdot\|_Y)$ be a Banach space (on $\mathbb{K}$). Then $\mathcal{L}(X, Y)$ is a Banach space.*

Proof We already know from Proposition 3.6 that $\mathcal{L}(X, Y)$ is a normed space, hence we are left to prove completeness. To this aim, we consider a Cauchy sequence $(T_n)_n$ in $\mathcal{L}(X, Y)$ and we show that there exists $T \in \mathcal{L}(X, Y)$ such that $\|T_n - T\| \to 0$ as $n \to +\infty$. To this aim, we notice that the real sequence $(\|T_n\|)_n$ is bounded, since

$$\big|\, \|T_n\| - \|T_m\| \,\big| \le \|T_n - T_m\| \to 0, \quad \text{as } n, m \to \infty.$$

Hence there exists $C > 0$ such that

$$\|T_n\| \le C, \quad \text{for all } n \in \mathbb{N}. \tag{3.5}$$

Next we consider, for every fixed $x \in X$ the sequence $(T_n x)_n$ in Y, and we prove that it is a Cauchy sequence:

$$\|T_n x - T_m x\|_Y = \|(T_n - T_m)x\|_Y \le \|T_n - T_m\|\,\|x\|_X \to 0,$$

since $(T_n)_n$ is a Cauchy sequence. From completeness of Y there exists $y \in Y$ such that $T_n x \to y$: we define the map T that acts on x as $Tx = y$, for every $x \in X$. We conclude proving that $T \in \mathcal{L}(X, Y)$ and that $\|T_n - T\| \to 0$, as $n \to +\infty$.

Linearity: take $x_1, x_2 \in X$,

$$T(x_1 + x_2) = \lim_{n\to\infty} T_n(x_1 + x_2) = \lim_{n\to\infty} T_n x_1 + \lim_{n\to\infty} T_n x_2 = T x_1 + T x_2.$$

Analogously $T(\alpha x) = \alpha(Tx)$, for all $x \in X$ and $\alpha \in \mathbb{K}$.

Continuity:

$$\|Tx\|_Y = \lim_{n\to\infty} \|T_n x\|_Y \le \limsup_{n\to\infty} \|T_n\|\,\|x\|_X \le C\,\|x\|_X,$$

where C given by (3.5). This proves that $T \in \mathcal{L}(X, Y)$.

Finally we show that $\|T_n - T\| \to 0$, as $n \to +\infty$. Take $x \in X$, $\|x\|_X = 1$:

$$\|T_n x - Tx\|_Y = \lim_{m\to\infty} \|T_n x - T_m x\|_Y \le \limsup_{m\to\infty} \|T_n - T_m\|\,\|x\|_X = \limsup_{n\to\infty} \|T_n - T_m\|,$$

hence

$$\|T_n - T\| = \sup_{\|x\|_X = 1} \|(T_n - T)x\| \leq \limsup_{m \to +\infty} \|T_n - T_m\|.$$

From definition of Cauchy sequence, for every $\varepsilon > 0$ there exists $N_\varepsilon \in \mathbb{N}$ such that $\|T_n - T_m\| \leq \varepsilon, \forall n, m \geq N_\varepsilon$. Thus $\|T_n - T\| \leq \varepsilon$, for all $n \geq N_\varepsilon$, proving convergence of T_n to T. $\qquad\square$

Corollary 3.11 *Let $(X, \|\cdot\|)$ be a normed space over $\mathbb{K}$. Then $X' = \mathcal{L}(X, \mathbb{K})$ is a Banach space. The norm is given by*

$$\|f\|_{X'} = \sup_{\|x\|_X = 1} |f(x)|, \quad f \in X'.$$

3.1 The Hahn-Banach Theorem

<u>Notation</u> Let $(X, \|\cdot\|)$ be a normed space over $\mathbb{K}$. Given $f \in X'$ we denote by

$$f(x) = \langle f, x \rangle_{X',X} = \langle f, x \rangle \tag{3.6}$$

the action of f on $x \in X$. As a consequence of previous arguments, we write

$$|\langle f, x \rangle_{X',X}| \leq \|f\|_{X'} \|x\|_X, \quad \forall f \in X', \forall x \in X.$$

Motivation: How Many Extensions of a Functional Are There?

Let $X = \mathbb{R}^n$ and $f \in X'$. As remarked in Example 3.2, there exists a unique vector $a_f \in \mathbb{R}^n$ such that

$$\langle f, x \rangle = a_f \cdot x, \quad \text{for every } x \in \mathbb{R}^n. \tag{3.7}$$

Consider instead $X = \mathbb{R}^3$ and $Y := \{(x_1, x_2, 0), x_i \in \mathbb{R}, i = 1, 2\}$. In general one can define the dual of a subspace $Y \subset X$, for any X normed space, as $Y' := \mathcal{L}(Y, \mathbb{K})$, since $(Y, \|\cdot\|_X)$ is itself a normed space. Take $g \in Y'$, then there exist $b_1, b_2 \in \mathbb{R}$ such that

$$\langle g, (x_1, x_2, 0) \rangle_{Y',Y} = b_1 x_1 + b_2 x_2.$$

Consider the vector $a = (b_1, b_2, 0) \in \mathbb{R}^3$ and $f_a \in X'$ given by $\langle f_a, x \rangle = a \cdot x$. Clearly $f_a \in X'$, and $f_a|_Y = g$. But actually if we consider any vector $\tilde{a} = (b_1, b_2, \gamma) \in \mathbb{R}^3$ and the corresponding element $f_{\tilde{a}}$ in X', $\langle f_{\tilde{a}}, x \rangle := \tilde{a} \cdot x$, we also obtain that $f_{\tilde{a}}|_Y = g$. Hence there exist infinitely many element of X' extending $g \in Y'$.

We can generalize this procedure: let X be a finite dimensional normed space and let $Y \subset X$ be a linear subspace of X. Then for any $g \in Y'$ there exist infinitely many $f \in X'$ such that $f|_Y \equiv g$.

Let us consider the same problem in an infinite dimensional setting: $X = \mathcal{C}(I)$, $I = [\alpha, \beta] \subset \mathbb{R}$. $Y = \mathcal{P}^k(I)$, subspace of X. Let $E = \{e_j, e_j(t) = t^j, \ j = 0, \ldots, k\}$, $Y = \mathrm{lsp}(E)$. For every $p \in \mathcal{P}^k(I)$ denote by $a_p \in \mathbb{R}^{k+1}$ the vector of coefficients of p. Given any $g \in Y'$ there exists a unique $b \in \mathbb{R}^{k+1}$ such that $\langle g, p \rangle_{Y',Y} = b \cdot a$. What can we say about extensions of g?

The Real Hahn-Banach Theorem

Theorem 3.12 (Hahn-Banach) *Let X be a vector space over $\mathbb{R}$ and $p : X \to \mathbb{R}$ a map such that*

$$p(tx + (1-t)y) \leq tp(x) + (1-t)p(y), \quad \forall x, y \in X, \ t \in [0, 1]. \qquad (3.8)$$

Let $Y \subset X$ be a subspace of X and $\lambda : Y \to \mathbb{R}$ be a linear function such that $\lambda(y) \leq p(y)$, $\forall y \in Y$. Then there exists a linear function $\Lambda : X \to \mathbb{R}$ such that

1. $\Lambda(x) \leq p(x), \quad \forall x \in X;$
2. $\Lambda(y) = \lambda(y), \quad \forall y \in Y.$

Before proving the Hahn-Banach Theorem, we anticipate the following corollary, which is an immediate important application to normed spaces.

Corollary 3.13 *Let $(X, \|\cdot\|)$ be a normed space over $\mathbb{R}$, $Y \subset X$ a subspace and $\lambda \in Y'$. Then there exists $\Lambda \in X'$ such that*

(i) $\Lambda|_Y \equiv \lambda;$
(ii) $\|\Lambda\|_{X'} = \|\lambda\|_{Y'}.$

Proof of Corollary 3.13. We apply the Hahn-Banach Theorem 3.12: for any $\alpha > 0$ the function $p : X \to \mathbb{R}$, $p(x) := \alpha \|x\|$ satisfies the convexity property (3.8). In particular, choosing $p(x) = \|\lambda\|_{Y'} \|x\|$ we have

$$\lambda(x) \leq |\lambda(x)| \leq \|\lambda\|_{Y'} \|y\|, \quad \forall y \in Y.$$

Thus, Theorem 3.12 ensures existence of $\Lambda \in X'$ extending λ and such that $\Lambda(x) \leq p(x) = \|\lambda\|'_Y \|x\|$, $\forall x \in X$. Finally, since from linearity

$$-\Lambda(x) = \Lambda(-x) \leq \|\lambda\|_{Y'} \|-x\| = \|\lambda\|_{Y'} \|x\|, \quad \forall x \in X,$$

we have that $\|\Lambda\|_{X'} = \|\lambda\|_{Y'}$. $\qquad\qquad\square$

Proof of the Hahn-Banach Theorem If $Y = X$ there is nothing to prove, thus, we suppose that there exists $z \in X \setminus Y$ and notice that, since Y is a subspace of X, $z \neq 0$. The proof is divided in two steps.

Step 1: we introduce the subspace $Y_z := Y + \mathbb{R}z = \{x = y + tz, y \in Y, t \in \mathbb{R}\}$ and we prove that there exists a map $\tilde{\lambda} : Y_z \to \mathbb{R}$ such that

(i) $\tilde{\lambda}$ is linear;
(ii) $\tilde{\lambda}(y) = \lambda(y), \quad \forall y \in Y;$

(iii) $\tilde\lambda(x) \le p(x), \quad \forall x \in Y_z$.

For every $x = y + tz \in Y_z$ we can write $\tilde\lambda(y + tz) = \tilde\lambda(y) + t\tilde\lambda(z) = \lambda(x) + t\tilde\lambda(z)$. Thus, proving the existence of $\tilde\lambda$ is equivalent to finding a value $\gamma := \tilde\lambda(z) \in \mathbb{R}$ such that (iii) holds. Equivalently we look for γ such that

$$\tilde\lambda(y + tz) = \lambda(y) + t\gamma \le p(y + tz), \quad \forall y \in Y, \ \forall t \in \mathbb{R}. \tag{3.9}$$

To this aim, we show that

$$\sup_{\substack{y \in Y \\ t \in \mathbb{R}^+}} \frac{1}{t}(\lambda(y) - p(y - tz)) \le \inf_{\substack{y \in Y \\ t \in \mathbb{R}^+}} \frac{1}{t}(-\lambda(y) + p(y + tz)). \tag{3.10}$$

If (3.10) holds, then there exists $\gamma \in \mathbb{R}$ such that

$$\frac{1}{t}(\lambda(y) - p(y - tz)) \le \gamma \le \frac{1}{t}(-\lambda(y) + p(y + tz)), \quad \forall y \in Y, \ \forall t \in \mathbb{R}^+.$$

These inequalities are equivalent to

$$\lambda(y) - t\gamma \le p(y - tz) \quad \text{and} \quad \lambda(y) + t\gamma \le p(y + tz), \quad \forall y \in Y, \ \forall t \in \mathbb{R}^+,$$

and since $\lambda(y) \le p(y), \forall y \in Y$ we have that

$$\lambda(y) + t\gamma \le p(y + tz), \quad \forall y \in Y, \ \forall t \in \mathbb{R},$$

concluding the proof of (3.9). Thus, we are left to prove (3.10) to conclude this first step. Let $y_1, y_2 \in Y$ and let $\alpha, \beta > 0$. Then

$$\begin{aligned}
\alpha\lambda(y_1) + \beta\lambda(y_2) = \lambda(\alpha y_1 + \beta y_2) &= \lambda\left((\alpha + \beta)\left[\frac{\alpha}{\alpha + \beta}y_1 + \frac{\beta}{\alpha + \beta}y_2\right]\right) \\
&= (\alpha + \beta)\lambda\left(\frac{\alpha}{\alpha + \beta}y_1 + \frac{\beta}{\alpha + \beta}y_2\right) \\
&\le (\alpha + \beta)p\left(\frac{\alpha}{\alpha + \beta}y_1 + \frac{\beta}{\alpha + \beta}y_2\right) \qquad (3.11) \\
&= (\alpha + \beta)p\left(\frac{\alpha}{\alpha + \beta}(y_1 - \beta z) + \frac{\beta}{\alpha + \beta}(y_2 + \alpha z)\right) \\
&\le \alpha p(y_1 - \beta z) + \beta p(y_2 + \alpha z).
\end{aligned}$$

This is equivalent to

$$\frac{1}{\beta}[\lambda(y_1) - p(y_1 - \beta z)] \leq \frac{1}{\alpha}[-\lambda(y_2) + p(y_2 + \alpha z)], \quad \forall y_1, y_2 \in Y, \ \forall \alpha, \beta > 0,$$

which is a different way to write (3.10).
Step 2: we introduce the set

$$\mathcal{E} := \{e : \mathrm{dom}(e) \to \mathbb{R}, \ \text{such that } \mathrm{dom}(e) \text{ is a subsapce of } X, \ Y \subseteq \mathrm{dom}(e), \tag{3.12}$$

$$e \text{ is linear}, \ e|_Y = \lambda, e(x) \leq p(x), \quad \forall x \in \mathrm{dom}(e)\}. \tag{3.13}$$

We notice that $\mathcal{E}$ is nonempty since $\lambda, \tilde{\lambda} \in \mathcal{E}$. We introduce the order relation in $\mathcal{E}$ given by: $e \subseteq f$ if $\mathrm{dom}(e) \subseteq \mathrm{dom}(f)$ and $f|_{\mathrm{dom}(e)} = e$. Let $\mathcal{U} = \{e_\alpha, \alpha \in \mathcal{A}\}$ be a totally ordered subsystem of $\mathcal{E}$, (i.e., for every $\alpha, \beta \in \mathcal{A}$ with $\alpha \neq \beta$, then either $e_\alpha \subseteq e_\beta$ or $e_\beta \subseteq e_\alpha$). We define

$$\mathrm{dom}(g) := \bigcup_{\alpha \in \mathcal{A}} \mathrm{dom}(e_\alpha), \quad \text{and } g(x) = e_{\overline{\alpha}}(x), \quad \forall x \in \mathrm{dom}(g),$$

where $x \in \mathrm{dom}(e_{\overline{\alpha}})$, and we remark that g is well defined. Clearly g is an upper bound of $\mathcal{U}$, thus, from Zorn's Lemma there exists a maximal element G of $\mathcal{E}$. We now show that $\mathrm{dom}(G) \equiv X$, so that, putting $\Lambda = G$ we have concluded the proof. Suppose by contradiction that $Y^* := \mathrm{dom}(G) \subset X$, and let $0 \neq z \in X \setminus Y^*$. Then repeating Step 1 on the subspace $Y^* + \mathbb{R}z := \{y + tz, \ y \in Y^*, t \in \mathbb{R}\}$, we obtain $\tilde{G} : Y^* + \mathbb{R}z \to \mathbb{R}$, satisfying properties $(i) - (iii)$. Clearly $\tilde{G} \in \mathcal{E}$ and $G \subsetneq \tilde{G}$, contradicting maximality of G. This concludes the proof. $\qquad \square$

The Complex Hahn-Banach Theorem

Lemma 3.14 *Let X be a vector space over $\mathbb{C}$, a map $F : X \to \mathbb{C}$ is linear if and only if it has the form*

$$F(x) = f(x) - if(ix), \quad \forall x \in X,$$

where $f : X \to \mathbb{R}$ is a linear map over $\mathbb{R}$.

Proof $\implies$ First assume that F is linear and write $F(x) = \alpha(x) + i\beta(x)$, $x \in X$, with $\alpha, \beta : X \to \mathbb{R}$ linear on $\mathbb{R}$. By linearity

$$F(ix) = iF(x) \quad \implies \quad \beta(x) = -\alpha(ix),$$

thus, $F(x) = \alpha(x) - i\alpha(ix), \forall x \in X$.

$\impliedby$ Next assume that $F(x) = f(x) - if(ix), \forall x \in X$. Clearly $\forall x_1, x_2 \in X$ we have $F(x_1 + x_2) = F(x_1) + F(x_2)$ and $\forall \lambda \in \mathbb{R}$ we have $F(\lambda x) = \lambda F(x)$ by linearity of f on $\mathbb{R}$. We are left to show that $F(\lambda x) = \lambda F(x)$, $\forall \lambda \in \mathbb{C}$. Thus, it is enough to prove it for $\lambda = i$:

$$F(ix) = f(ix) - if(-x) = f(ix) + if(x), \quad \text{and} \quad iF(x) = if(x) + f(ix),$$

concluding the proof. $\qquad\square$

Theorem 3.15 (The complex Hahn-Banach) *Let X be a linear space over $\mathbb{C}$ and $p : X \to \mathbb{R}$ a map such that $p(\alpha x + \beta y) \leq |\alpha| p(x) + |\beta| p(y)$, $\forall x, y \in X$, $\forall \alpha, \beta \in \mathbb{C}$ such that $|\alpha| + |\beta| = 1$. Let $Y \subseteq X$ be a subspace of X and $\lambda : Y \to \mathbb{C}$ a linear function such that $|\lambda(y)| \leq p(y)$, $\forall y \in Y$. Then there exists a linear function $\Lambda : X \to \mathbb{C}$ such that*

(i) $|\Lambda(x)| \leq p(x), \quad \forall x \in X$;
(ii) $\Lambda(y) = \lambda(y), \quad \forall y \in Y$.

Proof First of all we observe that $p(\alpha x) \leq |\alpha| p(x)$, $\forall x \in X$, $\forall \alpha \in \mathbb{C}$, such that $|\alpha| = 1$. From Lemma 3.14 we also have that there exists $\ell : Y \to \mathbb{R}$ linear on $\mathbb{R}$ such that $\lambda(y) = \ell(y) - i\ell(iy)$, $\forall y \in Y$. In addition $\ell(y) \leq |\lambda(y)| \leq p(y)$, $\forall y \in Y$. Hence we apply the real Hahn-Banach Theorem 3.12 to ℓ obtaining that there exists $L : X \to \mathbb{R}$ linear on $\mathbb{R}$ and such that

(i) $L(x) \leq p(x), \quad \forall x \in X$;
(ii) $L(y) = \ell(y), \quad \forall y \in Y$.

We define $\Lambda(x) := L(x) - iL(ix)$, $\forall x \in X$. We are left to prove that $|\Lambda(x)| \leq p(x)$, $\forall x \in X$. Given $x \in X$ we may write $\Lambda(x) = |\Lambda(x)| e^{i\theta}$ for some $\theta \in [0, 2\pi]$. Thus

$$|\Lambda(x)| = \Lambda(x) e^{-i\theta} = \Lambda(e^{-i\theta} x) = L(e^{-i\theta} x) - iL(ie^{-i\theta} x).$$

Noticing that $|\Lambda(x)|$, $L(e^{-i\theta} x)$ and $L(ie^{-i\theta} x)$ are reals, we must have $L(ie^{-i\theta} x) = 0$ and thus,

$$|\Lambda(x)| = L(e^{-i\theta} x) \leq p(e^{-i\theta} x) \leq p(x).$$

$\qquad\square$

The following is the analogous of Corollary 3.13 for the complex case.

Corollary 3.16 *Let $(X, \|\cdot\|)$ be a normed space over $\mathbb{K}$ and let $Y \subseteq X$ be a subspace. Let $\lambda \in Y'$, then there exists $\Lambda \in X'$ such that*

(i) $\|\Lambda\|_{X'} = \|\lambda\|_{Y'}$;
(ii) $\Lambda_Y = \lambda$.

3.2 Applications of the Hahn-Banach Theorem

First Applications

Proposition 3.17 *Let $(X, \|\cdot\|)$ be a normed space over $\mathbb{K}$ and let $x_0 \in X$, $x_0 \neq 0$. There exist $\lambda, \tilde{\lambda} \in X'$ such that*

1. $\langle \lambda, x_0 \rangle = \|x_0\|, \quad \|\lambda\|_{X'} = 1;$
2. $\left\langle \tilde{\lambda}, x_0 \right\rangle = 1.$

Proof Define the subspace $Y := \{tx_0, \ t \in \mathbb{K}\}$ and the map $g : Y \to \mathbb{K}$, given by $g(y) = g(tx_0) = t \|x_0\|$. Clearly g is linear and $|g(y)| = |t| \|x_0\| = \|y\|$, $\forall y \in Y$, hence $g \in Y'$ and $\|g\|_{Y'} = 1$. Thus, using Corollary 3.13 we find $G \in X'$ such that $\|G\|_{X'} = 1$ and $G|_Y = g$. In particular $G(x_0) = g(x_0) = \|x_0\|$. Hence we set $\lambda = G$.

Analogously we find $\tilde{\lambda}$ putting $\tilde{g} : Y \to \mathbb{K}$, $\tilde{g}(tx_0) = t$, and repeating the same proof. $\qquad\square$

Corollary 3.18 *Let $(X, \|\cdot\|)$ be a normed space over $\mathbb{K}$ and let $x_0 \in X$ such that $\langle \lambda, x_0 \rangle = 0$, $\forall \lambda \in X'$. Then $x_0 = 0$.*

Definition 3.19 Let $(X, \|\cdot\|)$ be a normed space, $z \in X$, $A \subset X$, $A \neq \emptyset$. We define

$$\mathrm{dist}(\cdot, A) : X \to [0, \infty[, \quad \mathrm{dist}(z, A) := \inf\{\|z - x\|, x \in A\}.$$

Proposition 3.20 *Let $(X, \|\cdot\|)$ be a normed space over $\mathbb{K}$ and let $Y \subseteq X$ a proper subspace. Take $z \in X \setminus Y$ a vector such that $\mathrm{dist}(z, Y) > 0$. Then there exists $\lambda \in X'$ such that*

1. $\langle \lambda, y \rangle = 0, \quad \forall y \in \overline{Y};$
2. $\langle \lambda, z \rangle = \mathrm{dist}(z, Y).$

Proof Introduce the subspace $Y_z := \{x = y + tz, \ t \in \mathbb{K}\}$ and the map $g : Y_z \to \mathbb{K}$ given by $g(x) = t \, \mathrm{dist}(z, Y)$. Since $g \in Y_z$, using the Hahn-Banach Theorem we find $\lambda \in X'$ satisfying the required properties. $\qquad\square$

Corollary 3.21 *Let $(X, \|\cdot\|)$ be a normed space over $\mathbb{K}$. Then $\|x\| = \sup_{\|\lambda\|_{X'}=1} |\langle \lambda, x \rangle|$, $\forall x \in X$.*

Proof We know that for all λ in X' and for all x in X, $|\langle \lambda, x \rangle| \leq \|\lambda\|_{X'} \|x\|$. Hence $\|x\| \geq \sup_{\|\lambda\|_{X'}} |\langle \lambda, x \rangle|$. But using Proposition 3.17 we also know that there exists $\lambda \in X'$, $\|\lambda\|_{X'} = 1$ such that $\langle \lambda, x \rangle = \|x\|$, this concludes the proof. $\qquad\square$

Reflexivity

Definition 3.22 Let $(X, \|\cdot\|_X)$ be a normed space and let $(X', \|\cdot\|_{X'})$ be its dual space (recall that X' is a Banach space). We may consider the dual space of X':

$$(X')' := \mathcal{L}(X', \mathbb{K}),$$

we call it the *bidual space* of X and denote it with X''.

Clearly X'' is a Banach space endowed with the norm

$$\|\xi\|_{X''} = \sup_{\|\lambda\|_{X'}=1} |\langle \xi, \lambda \rangle_{X'',X'}|.$$

Define the map $J : X \to X''$ as follows: $\langle J(x), \lambda \rangle_{X'',X'} = \langle \lambda, x \rangle_{X',X}$, for every $\lambda \in X'$ and $x \in X$. Clearly $J : X \to X''$ is a linear map, we remark that it is also an isometry, i.e., $\|J(x)\|_{X''} = \|x\|_X$, for all elements x in X. Indeed for every fixed x in X the map $g : X' \to \mathbb{K}$, given by $g(\lambda) = \langle \lambda, x \rangle_{X',X}$, satisfies

$$|g(\lambda)| \leq \|\lambda\|_{X'} \|x\|_X, \qquad \forall \lambda \in X',$$

since x is fixed. Hence g is an element of the bidual space X'', and $\|g\|_{X''} \leq \|x\|_{X'}$. Actually we have equality, since $\|g\|_{X''} = \sup_{\|\lambda\|_{X'}=1} |\langle \lambda, x \rangle_{X',X}| = \|x\|_X$. Thus, $\|J(x)\|_{X''} = \|x\|_X$, for all x in X. The map J defines a subspace $J(X)$ of X''. We give the following definition:

Definition 3.23 Using previous notation we say that the space X is *reflexive* if $J(X) = X''$.

Convexity

Definition 3.24 Let X be a linear space and $C \subseteq X, C \neq \emptyset$. We say that C is *convex* if for any pair of points x, y in C and for any t in $[0, 1]$, the point $(1 - t)x + ty$ still belongs to C.

Definition 3.25 Let $(X, \|\cdot\|)$ be a normed space over $\mathbb{R}$.

1. Given $\lambda \in X', \lambda \neq 0$, and $\alpha \in \mathbb{R}$, we call *hyperplane of equation* $\lambda = \alpha$, the set $\{\lambda = \alpha\} := \{x \in X : \langle \lambda, x \rangle = \alpha\}$.
2. Given $A, B \subseteq X$, nonempty subsets, $\lambda \in X'$, $\lambda \neq 0$ and $\alpha \in \mathbb{R}$, we say that $\{\lambda = \alpha\}$ separates (weakly) A and B if

$$\langle \lambda, x \rangle \leq \alpha, \quad \forall x \in A,$$

and

$$\langle \lambda, y \rangle \geq \alpha, \quad \forall y \in B.$$

3. In the same notations we say that $\{\lambda = \alpha\}$ separates strongly (or strictly) A and
   B if there exists a positive ε such that

$$\langle \lambda, x \rangle \leq \alpha - \varepsilon, \quad \forall x \in A,$$

and

$$\langle \lambda, y \rangle \geq \alpha + \varepsilon, \quad \forall y \in B.$$

Definition 3.26 (*The Minkowski functional or gauge*) Let $(X, \|\cdot\|)$ be a normed
space over $\mathbb{R}$ and $C \subseteq X$ an open convex subset containing 0. We set

$$\mathrm{p}(x) := \inf\{\alpha > 0 : \frac{1}{\alpha}x \in C\}, \quad x \in X.$$

$\mathrm{p} : X \to [0, \infty[$ is called the *Minkowski functional* or gauge associated to C.

Exercise: Let $C := B(0, 1)$. Then $\mathrm{p}(x) = \|x\|$.

Proposition 3.27 (Properties of p) *The following properties hold:*

1. $\mathrm{p}(\lambda x) = \lambda \mathrm{p}(x), \quad \forall x \in X, \forall \lambda > 0$;
2. *there exists* $M \geq 0$ *such that* $\mathrm{p}(x) \leq M \|x\|, \quad \forall x \in X$;
3. $C = \{x \in X : \mathrm{p}(x) < 1\}$;
4. $\mathrm{p}(x + y) \leq \mathrm{p}(x) + \mathrm{p}(y), \quad \forall x, y \in X$.

Proof 1. If $\lambda = 0$ then trivially $\mathrm{p}(0) = 0$. If instead $\lambda > 0$ then

$$\mathrm{p}(\lambda x) = \inf\{\alpha > 0 : \frac{1}{\alpha}\lambda x \in C\} \overset{\beta := \frac{\alpha}{\lambda}}{=} \inf\{\lambda \beta : \frac{1}{\beta}x \in C\} = \lambda \inf\{\beta : \frac{1}{\beta}x \in C\} = \lambda \mathrm{p}(x).$$

2. $0 \in C$, and C is open, thus, there exists $r > 0$ such that $\overline{B(0, r)} \subseteq C$. Take any
   $x \in X$ and note that

$$\frac{1}{\frac{\|x\|}{r}}x = r\frac{x}{\|x\|} \in \overline{B(0, r)} \subseteq C.$$

Consequently $\mathrm{p}(x) \leq \frac{1}{r}\|x\|$, and putting $M = \frac{1}{r}$ this proves point 2.
3. If $x \in C$, then $\mathrm{p}(x) < 1$: indeed, since C is open, there exists $\varepsilon > 0$ such that
   $(1 + \varepsilon)x \in C$. Hence $\mathrm{p}(x) \leq \frac{1}{1+\varepsilon} < 1$. Conversely if $\mathrm{p}(x) < 1$, then $\exists \lambda \in]0, 1[$
   such that $\frac{1}{\lambda}x \in C$. Thus, we write

$$x = \lambda\left(\frac{x}{\lambda}\right) + (1 - \lambda)0 \in C.$$

4. Let $x, y \in X$ and take any $\varepsilon > 0$. Consider the vectors $\frac{1}{\mathrm{p}(x)+\varepsilon}x$ and $\frac{1}{\mathrm{p}(x)+\varepsilon}y$.
   From point 1. we know that

$$p\left(\frac{1}{p(x)+\varepsilon}x\right) = \frac{1}{p(x)+\varepsilon}p(x) < 1, \quad \text{and} \quad p\left(\frac{1}{p(y)+\varepsilon}y\right) = \frac{1}{p(y)+\varepsilon}p(y) < 1,$$

thus, from point 3 both vectors belong to C. Taking a convex combination of the two we get

$$\frac{p(x)+\varepsilon}{p(x)+p(y)+2\varepsilon}\frac{x}{p(x)+\varepsilon} + \frac{p(y)+\varepsilon}{p(x)+p(y)+2\varepsilon}\frac{y}{p(y)+\varepsilon} \in C,$$

and thus $\frac{1}{p(x)+p(y)+2\varepsilon}(x+y) \in C$. Hence, using point 3, we have that

$$p\left(\frac{1}{p(x)+p(y)+2\varepsilon}(x+y)\right) = \frac{p(x+y)}{p(x)+p(y)+2\varepsilon} \leq 1 \implies p(x+y) \leq p(x)+p(y)+2\varepsilon.$$

By the arbitrariness of $\varepsilon > 0$ this concludes the proof. $\qquad\square$

Proposition 3.28 *Let $(X, \|\cdot\|)$ be a normed space over $\mathbb{R}$, let $C \subseteq X$ open, convex and nonempty. Let $x_0 \notin C$, then there exists $\lambda \in X'$, $\lambda \neq 0$, such that $\langle \lambda, x \rangle \leq \langle \lambda, x_0 \rangle$, $\forall x \in C$. In other words, the hyperplane $\{\lambda = \langle \lambda, x_0 \rangle\}$ separates (weakly) the sets C and $\{x_0\}$.*

Proof Up to translation we may suppose that $0 \in C$, so that we can define the gauge map p of C (see Definition 3.26). Then introduce the space $Y := \{y = tx_0, \, t \in \mathbb{R}\} \subseteq X$, and remark that $x_0 \neq 0$. Define $g : Y \to \mathbb{R}$, $g(tx_0) := t$, which is linear and such that $g(x_0) = 1$. Moreover,

$$g(tx_0) = t \leq p(tx_0), \; \forall t \leq 0 \quad \text{and} \quad g(tx_0) = t \leq tp(x_0) = p(tx_0), \quad \forall t > 0,$$

since $x_0 \notin C$. Hence $g(y) \leq p(y)$ for all $y \in Y$. Thus, we apply the Hahn-Banach Theorem to g obtaining the functional $G : X \to \mathbb{R}$, linear and such that $G|_Y = g$ and $G(x) \leq p(x), \forall x \in X$. In addition $G(x) \leq p(x) \leq M \|x\|, \forall x \in X$, so that $G \in X'$. Finally

$$G(x) \leq p(x) < 1, \quad \forall x \in C, \quad \text{and} \quad G(x_0) = g(x_0) = 1.$$

Thus, setting $\lambda = G$ we have $\lambda \in X'$ and $\langle \lambda, x \rangle \leq \langle \lambda, x_0 \rangle, \forall x \in C$. $\qquad\square$

Geometric Forms of the Hahn-Banach Theorem

Theorem 3.29 (1st geometric form of Hahn-Banach) *Let $(X, \|\cdot\|)$ be a normed space over $\mathbb{R}$ and $A, B \subset X$ be two convex subsets nonempty and disjoint. Suppose that A is open. Then there exists $\lambda \in X'$, $\lambda \neq 0$, and $\alpha \in \mathbb{R}$ such that the hyperplane $\{\lambda = \alpha\}$ separates A and B in weak sense.*

Proof Consider the set

$$C = A - B := \{z = x - y, \, x \in A, \, y \in B\}.$$

One can easily check that C is nonempty, convex and that $0 \notin C$. We also show that C is open: indeed for any $y \in B$ the set $A_y := \{z = x - y,\ x \in A\}$ is open and $C = \bigcup_{y \in B} A_y$, hence it is open. Thus, we can apply Proposition 3.28 with $x_0 = 0$, obtaining that $\{\lambda = 0\}$ separates C and $\{0\}$, i.e.,

$$\langle \lambda, z \rangle \leq \langle \lambda, 0 \rangle = 0 \quad \Longrightarrow \quad \langle \lambda, x \rangle \leq \langle \lambda, y \rangle, \quad \forall x \in A,\ y \in B,$$

which concludes the proof. $\qquad\qquad\qquad\qquad\qquad\qquad\qquad\qquad\qquad\qquad\qquad\quad$ $\square$

Theorem 3.30 (2nd geometric form of the Hahn-Banach Theorem) *Let $(X, \|\cdot\|)$ be a normed space over $\mathbb{R}$ and let $A, B \subset X$ be two convex nonempty disjoint subsets. Suppose that A is closed and B is compact. Then there exists $\lambda \in X'$, $\lambda \neq 0$, and $\alpha \in \mathbb{R}$ such that the hyperplane $\{\lambda = \alpha\}$ separates A and B in strong sense.*

Proof For any $r > 0$ define the open and convex sets

$$A^r = A + B(0, r) := \{x + \xi,\ x \in A,\ |\xi| < r\}, \quad B^r = B + B(0, r) := \{y + \xi,\ y \in B,\ |\xi| < r\}.$$

We claim that there exists $\varepsilon > 0$, sufficiently small, such that $A^\varepsilon \cap B^\varepsilon = \emptyset$. If the claim holds, we can apply the first geometric form of the Hahn-Banach Theorem 3.29 to the pair $\{A^\varepsilon, B^\varepsilon\}$, obtaining the existence of a hyperplane $\{\lambda, \alpha\}$ such that

$$\langle \lambda, x \rangle \leq \langle \lambda, y \rangle,\ \forall x \in A^\varepsilon, y \in B^\varepsilon,$$

which, expanding the definitions, reads

$$\langle \lambda, x + \xi \rangle \leq \langle \lambda, y + \eta \rangle,\ \forall x \in A, y \in B, \forall |\xi|, |\eta| < \varepsilon.$$

Thus, choosing ξ such that $\langle \lambda, \xi \rangle = \gamma > 0$, and setting $\eta = -\xi$, so that

$$\langle \lambda, x \rangle + \gamma \leq \langle \lambda, y \rangle - \gamma, \quad \forall x \in A,\ \text{and}\ \forall y \in B,$$

we conclude the proof. Thus, we are left to prove the claim. Suppose by contradiction that for all natural n there exists an element x_n in $A^{\frac{1}{n}} \cap B^{\frac{1}{n}}$. First we write $x_n = y_n + \xi_n$, with y_n in B and $\|\xi_n\| \leq \frac{1}{n}$. Next, since B is compact, there exists a subsequence $(y_{n_k})_k$ such that $y_{n_k} \overset{k \to \infty}{\to} y \in B$. But for all k we have that $x_{n_k} \in A^{\frac{1}{n_k}}$, and then

$$x_{n_k} = y_{n_k} + \xi_{n_k} \overset{k \to \infty}{\to} y \in \bigcap_{k \in \mathbb{N}} A^{\frac{1}{n_k}} = A.$$

Hence we obtain a contradiction since $y \in A \cap B = \emptyset$. $\qquad\qquad\qquad\qquad\qquad\quad$ $\square$

Further Consequences of the Hahn-Banach Theorem

Separability

Recall the definition of separable space in 2.6. We have the following result:

Theorem 3.31 *Let $(X, \|\cdot\|)$ be a normed space. If X' is separable, then X is separable.*

Proof Let $(\lambda_n)_n$ be a dense sequence in X'. Then, for all natural numbers n there exists x_n in X, $\|x_n\| = 1$, such that $\langle \lambda_n, x_n \rangle \geq \frac{1}{2} \|\lambda_n\|_{X'}$. We introduce the subspace $Y = \mathrm{lsp}(\{x_n, \ n \in \mathbb{N}\}) \subseteq X$ and we show that Y is dense in X. This concludes the proof since defining $\tilde{Y} := \mathrm{lsp}_{\mathbb{Q}}(\{x_n, n \in \mathbb{N}\})$, which is dense in Y and is also countable, we obtain separability of X. To prove that Y is dense we proceed by contradiction: suppose that there exists $z \in X$ and $\alpha > 0$ such that $\|z - y\| \geq \alpha$, for all y in Y. Applying Proposition 3.20 there exists $\tilde{\lambda} \in X'$, $\tilde{\lambda} \neq 0$ such that

$$\left\langle \tilde{\lambda}, y \right\rangle = 0 \ \forall y \in Y, \quad \text{and} \quad \left\langle \tilde{\lambda}, z \right\rangle > 0.$$

Take $(\lambda_{n_k})_k$ subsequence such that $\lambda_{n_k} \overset{k \to \infty}{\to} \tilde{\lambda} \in X'$. In particular $\|\lambda_{n_k}\|_{X'} \to \|\tilde{\lambda}\|_X'$. Recalling that, for all k, $\|x_{n_k}\| = 1$, and that $x_{n_k} \in Y$ (hence $\langle \tilde{\lambda}, x_{n_k} \rangle = 0$), we obtain:

$$\left\| \lambda_{n_k} - \tilde{\lambda} \right\|_{X'} \geq \left\langle \lambda_{n_k} - \lambda, \tilde{x}_{n_k} \right\rangle = \langle \lambda_{n_k}, x_{n_k} \rangle \geq \frac{1}{2} \|\lambda_{n_k}\|_{X'}.$$

Hence $\|\lambda_{n_k}\|_{X'} \leq 2 \left\| \lambda_{n_k} - \tilde{\lambda} \right\|_{X'} \to 0$ and since $\|\lambda_{n_k}\|_{X'} \to \|\tilde{\lambda}\|_{X'}$ we have $\tilde{\lambda} = 0$, which is a contradiction. Hence $Y = \mathrm{lsp}(\{x_n, n \in \mathbb{N}\})$ is dense in X. $\qquad\square$

Differentiable Functionals

We now conclude this section with the last example of application of the Hahn-Banach Theorem. Consider a continuous functional $u : \overline{I} \to X$, where I is an open real interval $I :=]\alpha, \beta[$. We say that u is differentiable at $t_0 \in I$ if there exists an element $u'(t_0)$ in X such that

$$\lim_{t \to t_0} \left\| \frac{u(t) - u(t_0)}{t - t_0} - u'(t_0) \right\|_X = 0.$$

We call $u'(t_0)$ derivative of u at the point t_0, remarking that if $u'(t_0)$ exists, it is unique. Analogously we can define $u'(\alpha)^+$ and $u'(\beta)^-$.

The problem is the following: suppose that $u : \overline{I} \to X$ is continuous and that u is differentiable at any point $t \in I$. If $u'(t) = 0$ for any t in I, can we say that $u(t) = u(\alpha)$ for all $t \in \overline{I}$?

We introduce the map $\phi : \overline{I} \to \mathbb{R}$ given by $\phi(t) = \langle \lambda, u(t) \rangle_{X',X}$, for all t in $\overline{I}$, where λ is an arbitrary fixed element of X'. Clearly ϕ is continuous, since it is the composition of continuous functions. If we consider the incremental quotient and use linearity, we obtain

$$\frac{\phi(t+h) - \phi(t)}{h} = \frac{\langle \lambda, u(t+h) \rangle - \langle \lambda, u(t) \rangle}{h} = \left\langle \lambda, \frac{u(t+h) - u(t)}{h} \right\rangle,$$

and from continuity of λ we have

$$\lim_{h \to 0} \frac{\phi(t+h) - \phi(t)}{h} = \langle \lambda, u'(t) \rangle = 0.$$

Hence $\phi'(t) = 0$ and thus, $\phi(t) = \phi(\alpha)$, for all t in I. From the expression of ϕ this means that

$$\langle \lambda, u(t) \rangle = \langle \lambda, u(\alpha) \rangle . \tag{3.14}$$

Since inequality (3.14) holds true for all $\lambda \in X'$ we have $u(t) = u(\alpha)$, for all t in the interval I.

3.3 The Baire Theorem

The second main result of this section is the Baire Theorem, that we now state and prove.

Definition 3.32 Let (M, d) be a metric space and let $S \subset M$. We say that S is *nowhere dense* if the interior part of the closure of S is empty.

Example 3.33 The following are examples of nowhere dense sets in finite dimensional spaces.

- Let $M = [0, 1] \subset \mathbb{R}$, with distance function defined as $\mathrm{d}(x, y) = |x - y|$, and consider the subset $S := \{\frac{1}{n}, \ n \in \mathbb{N}\}$. It is nowhere dense, since $\overline{S} = S \cup \{0\}$, which has empty interior.
- $M = \mathbb{R}^2$, $\mathrm{d}(x, y) = |x - y|$, then any finite family of lines is nowhere dense;
- $M = \mathbb{R}^3$, $\mathrm{d}(x, y) = |x - y|$, analogously any finite family of planes is nowhere dense.

The intuitive idea of these examples is that, as one has clearly in mind, it is not possible to cover $\mathbb{R}^2$ with a countable number of lines and analogously $\mathbb{R}^3$ with a countable number of planes. The Baire Theorem generalizes this concept to the non trivial infinite dimensional setting. We anticipate here that the proof is a contradiction argument, that one can visualize in the finite dimensional setup too.

Theorem 3.34 (Baire) *A complete metric space cannot be a countable union of nowhere dense sets.*

Proof Assume by contradiction that the complete metric space (M, d) can be written as

$$M = \bigcup_{n=1}^{\infty} A_n, \quad \text{with} \quad \mathrm{int}(\overline{A}_n) = \emptyset, \quad \forall n \in \mathbb{N}.$$

Clearly there exists an element $x_1 \in M \setminus \overline{A}_1$: otherwise we would have $M = \overline{A}_1$ and $\mathrm{int}\left(\overline{A}_1\right) \neq \emptyset$. Since $\overline{A}_1$ is closed, the distance between $\overline{A}_1$ and x_1 is strictly positive (recall the definition of distance from a set given in 3.19), and in particular there exist $r_1 > 0, d_1 > 0$ such that

$$0 < r_1 < \frac{1}{2} \quad B_1 := B(x_1, r_1) \subset M \setminus \overline{A}_1, \quad \text{and} \quad \mathrm{dist}\left(z, \overline{A}_1\right) \geq d_1 > 0, \quad \forall z \in B_1.$$

We repeat the argument in $\overline{B}_1$: there exists $x_2 \in \overline{B}_1 \setminus \overline{A}_2$, otherwise we would have $B_1 \subset \overline{A}_2$, but $\overline{A}_2$ has empty interior, thus, in particular it cannot contain the open B_1. Since $\overline{A}_2$ is closed, the distance between $\overline{A}_2$ and x_2 is strictly positive, and in particular there exist $r_2 > 0, d_2 > 0 \; 0 < r_2 < \frac{1}{4}$, such that

$$B_2 := B(x_2, r_2) \subseteq \overline{B}_1 \setminus \overline{A}_2, \quad \text{and} \quad \mathrm{dist}\left(z, \overline{A}_j\right) \geq d_j > 0, \quad \forall z \in B_j, \; j = 1, 2.$$

Iterating: for every $n \in \mathbb{N}^+$ there exists $x_n \in \overline{B}_{n-1} \setminus \overline{A}_n$, and there exist $r_n, d_n > 0$, $0 < r_n < 2^{-n}$, such that

$$B_n := B(x_n, r_n) \subseteq \overline{B}_{n-1} \setminus \overline{A}_n \quad \text{and} \quad \mathrm{dist}\left(z, \overline{A}_j\right) \geq d_j > 0 \quad \forall z \in B_j, \; j = 1, \ldots, n.$$

In particular we have

$$\mathrm{dist}\left(x_j, \overline{A}_j\right) \geq d_j \quad \forall j = 1, \ldots, n.$$

Now we remark that the sequence $(x_n)_n$ is a Cauchy sequence, indeed:

$$d\left(x_{n+j}, x_n\right) \leq \frac{1}{2^n} \quad \forall j, n \in \mathbb{N}.$$

Hence there exists $\tilde{x} \in M$ such that $(x_n)_n$ converges to $\tilde{x}$. But $\tilde{x}$ belongs to M which is the countable union of nowhere dense sets $(A_n)_n$, thus, in particular $\tilde{x}$ is contained in one of such sets, say $\tilde{x} \in A_k$. Thus

$$\mathrm{d}(x_n, \tilde{x}) \to 0, \quad \text{as } n \to +\infty, \quad \text{whilst} \quad \mathrm{dist}\left(x_n, \overline{A}_k\right) \geq d_k \geq 0 \quad \forall n \geq k,$$

giving a contradiction which concludes the proof. $\qquad \square$

Corollary 3.35 *Let (M, d) be a complete metric space and let $\{A_n, n \in \mathbb{N}\}$ be a countable family of open and dense subsets of M. Then*

$$A = \bigcap_{n \in \mathbb{N}} A_n \text{ is dense in } M.$$

Proof Let $C_n := A_n^c = M \setminus A_n$. For every n, the set C_n is closed and nowhere dense, being the complement of a dense set. Suppose by contradiction that $B(x, r) \subseteq A^c$, then

$$\overline{B(x, \frac{r}{2})} = \bigcup_{n \in \mathbb{N}} \overline{C_n \cap B(x, \frac{r}{2})},$$

which contradicts the Baire Theorem since $\overline{B(x, \frac{r}{2})}$ is complete. $\square$

3.4 Consequences of the Baire Theorem

We now prove many fundamental results, that come as applications of the Baire Theorem.

The following Banach-Steinhaus Theorem is often called in the literature *uniform boundedness principle*.

Theorem 3.36 (Banach-Steinhaus) *Let* $(X, \|\cdot\|_X)$ *be a Banach space and* $(Y, \|\cdot\|_Y)$ *a normed space. Let* $\mathcal{F} \subseteq \mathcal{L}(X, Y)$ *be a family with the following property:*

$$\forall x \in X, \ \exists M_x \geq 0, \ \text{such that} \ \|Tx\|_Y \leq M_x, \ \forall T \in \mathcal{F};$$

in other words, the set $\{Tx, \ T \in \mathcal{F}\}$ *is bounded in* Y *for every* x *in* X. *Then there exists* $M \geq 0$ *such that*

$$\|T\|_{\mathcal{L}(X,Y)} \leq M, \quad \forall T \in \mathcal{F}.$$

Proof For every $n \in \mathbb{N}$ we define

$$A_n := \{x \in X \ : \ \|Tx\|_Y \leq n, \ \forall T \in \mathcal{F}\} = \bigcap_{T \in \mathcal{F}} \{x \in X : \|Tx\|_Y \leq n\}.$$

We notice that A_n is closed for any n, since it is the intersection of closed sets. Given any element x of X, choosing $n > M_x$ we see that x belongs to A_n. Hence $X = \bigcup_{n=1}^{\infty} A_n$. From the Baire Theorem there exists $\bar{n} \in \mathbb{N}$ such that $\text{int}(\overline{A_{\bar{n}}}) \neq \emptyset$, implying that there exists a point x_0 in X and $\varepsilon > 0$ such that

$$\overline{B}(x_0, \varepsilon) = \{x \in X \ : \ \|x - x_0\|_X \leq \varepsilon\} \subseteq A_{\bar{n}}.$$

Hence $\|Tx\|_Y \leq \bar{n}$, for all x in $\overline{B}(x_0, \varepsilon)$ and for all $T \in \mathcal{F}$. Equivalently,

$$\|T(x_0 + \varepsilon\xi)\|_Y \leq \bar{n}, \quad \forall \|\xi\|_X \leq 1, \ \forall T \in \mathcal{F}.$$

Thus, we write

$$T(\xi) = \frac{1}{\varepsilon}T(x_0 + \varepsilon\xi) - \frac{1}{\varepsilon}Tx_0,$$

and we deduce that

$$\|T\xi\|_Y \le \frac{1}{\varepsilon}\|Tx_0\|_Y + \frac{1}{\varepsilon}\|T(x_0 + \varepsilon\xi)\|_Y \le \frac{1}{\varepsilon}M_{x_0} + \frac{1}{\varepsilon}\overline{n} = \frac{M_{x_0} + \overline{n}}{\varepsilon}.$$

Since this holds for every ξ with $\|\xi\|_X \le 1$, and for every $T \in \mathcal{F}$, taking the supremum on $\|\xi\|_X = 1$ we obtain

$$\|T\|_{\mathcal{L}(X,Y)} \le M, \quad \forall T \in \mathcal{F},$$

where we have set $M := \frac{M_{x_0} + \overline{n}}{\varepsilon}$, which concludes the proof. $\qquad\square$

Let $T \in \mathcal{L}(\mathbb{R}^n) = \mathcal{L}(\mathbb{R}^n, \mathbb{R}^n)$, there exists a real $n \times n$ matrix A such that

$$Tx = Ax \quad \forall x \in \mathbb{R}^n.$$

Thus, T is surjective $\Leftrightarrow$ A is invertible, (i.e., $\det A \ne 0$). In such case we have also T injective and hence invertible. Denoting by $S \in \mathcal{L}(\mathbb{R}^n)$ the inverse operator of T, $S = T^{-1}$ we have that $Sx = A^{-1}x$, for every element x of $\mathbb{R}^n$. This implies that for every open subset U of $\mathbb{R}^n$ we have

$$T(U) = S^{-1}(U) = \{y \in \mathbb{R}^n : y = Sx, \ x \in \mathbb{R}^n\},$$

hence $T(U)$ is open, since S is continuous. We now generalize this kind of property to map between general Banach spaces.

Theorem 3.37 (Open mapping) *Let $(X, \|\cdot\|_X)$ and $(Y, \|\cdot\|_Y)$ be Banach spaces over $\mathbb{K}$ and let $T \in \mathcal{L}(X, Y)$ be surjective. Then for every $U \subseteq X$ open, $T(U)$ is open in Y. We thus say that the map T is open.*

We set some notations used in the proof.
<u>Notations:</u> given $x \in X, \lambda \in \mathbb{K}, A \subseteq X$ and $r > 0$ we write:

$$x + A = \{z = x + y, \quad y \in A\};$$
$$\lambda A = \{z = \lambda y, \quad y \in A\};$$
$$B_r^X = B^X(0, r) = \{x \in x : \|x\|_X < r\};$$
$$B_r^X = r B_1^X;$$
$$T(x + \lambda A) = Tx + \lambda T(A);$$
$$T(B_r^X) = T(r B_1^X) = r T(B_1^X).$$

Proof To prove that the map T is open we have to prove the following claim: for every $x \in X$ and for any neighbourhood $\mathcal{U}$ of 0 in X the set $T(x + \mathcal{U})$ is a neighbourhood of Tx in Y.

By linearity we can consider only the case $x = 0$ and assume that $\mathcal{U}$ is an open ball: thus, we must show that for every positive radius r, there exists a corresponding ρ such that $B_\rho^Y \subseteq T(B_r^X)$, or equivalently $\rho B_1^Y \subseteq rT(B_1^X)$. Renaming $r = \frac{\rho}{r}$, the claim thus becomes: there exists a positive radius r such that

$$r B_1^Y \subseteq T(B_1^X).$$

In order to prove the claim we start using surjectivity of T to write $Y = \bigcup_{n=1}^\infty T(B_n^X) = \bigcup_{n=1}^\infty \overline{T(B_n^X)}$, and using the Baire Theorem we know that there exists at least a natural n_0 such that the interior part of the closure of $T(B_{n_0}^X)$ is nonempty. Since $\overline{T(B_{n_0}^X)} = n_0 \overline{T(B_1^X)}$, we deduce that there exists an open subset $\mathcal{W}$ of Y, contained in $\overline{T(B_1^X)}$. In particular $\mathcal{W}$ contains a ball, hence there exists $y_0 \in Y$ and a positive δ such that

$$B^Y(y_0, \delta) \subseteq \overline{T(B_1^X)}.$$

Using again surjectivity of T we denote by x_0 a preimage of y_0, i.e., $y_0 = Tx_0$. Thus, we can rewrite $B^Y(Tx_0, \delta) \subseteq \overline{T(B_1^X)}$. Using the notation introduced just before this proof, we write $B^Y(Tx_0, \delta) = Tx_0 + B_\delta^Y$, thus

$$B_\delta^Y \subseteq -Tx_0 + \overline{T(B_1^X)}.$$

Consider now the set

$$-Tx_0 + T(B_1^X) = \{T(x - x_0),\ x \in B_1^X\} \subseteq T(B_\alpha^X),$$

where α is a positive value such that $\alpha > \|x_0\| + 1$. We obtain $B_\delta^Y \subseteq \overline{T(B_\alpha^X)}$. Setting $\varepsilon := \frac{\delta}{\alpha}$ we have $B_\varepsilon^Y \subseteq \overline{T(B_1^X)}$. If we show that

$$\overline{T(B_1^X)} \subseteq T(B_2^X), \tag{3.15}$$

we obtain that $B_{\frac{\varepsilon}{2}}^Y \subseteq T(B_1^X)$, concluding the proof. Thus, we are left to show (3.15). Take $y \in \overline{T(B_1^X)}$ arbitrary, by definition of the closure, there exists an element $x_1 \in B_1^X$ such that $\|y - Tx_1\|_Y \leq \frac{\varepsilon}{2}$, meaning that $y - Tx_1 \in B_{\frac{\varepsilon}{2}}^Y \subseteq \overline{T(B_{\frac{1}{2}}^X)}$. Hence, there exists x_2 in $B_{\frac{1}{2}}^X$ such that

$$\|(y - Tx_1) - Tx_2\|_Y = \|y - Tx_1 - Tx_2\|_Y < \frac{\varepsilon}{4}, \quad \implies \quad y - T(x_1 + x_2) \in B_{\frac{\varepsilon}{4}}^Y \subseteq \overline{T(B_{\frac{1}{4}}^X)}.$$

Iterating we construct a sequence $x_n \in B^X_{\frac{1}{2^{n-1}}}$ such that $\|y - T(x_1 + \ldots + x_n)\|_Y < \frac{\varepsilon}{2^n}$. Consider now the series $\sum_{n=1}^{\infty} x_n$, we have

$$\left\| \sum_{n=N}^{N+P} x_n \right\|_X \leq \sum_{n=N}^{\infty} \|x_n\|_X < \sum_{n=N}^{\infty} \frac{1}{2^{n-1}} \to 0, \quad \text{as } N \to +\infty.$$

Hence the series converges to a point x. Estimating its norm we have

$$\|x\|_X \leq \sum_{n=1}^{\infty} \|x_n\|_X < \sum_{n=1}^{\infty} \frac{1}{2^{n-1}} = 2.$$

Thus, $x \in B^X_2$ and since

$$\left\| y - T(\sum_{n=1}^{N} x_n) \right\|_Y \leq \frac{\varepsilon}{2^N} \to 0, \quad \text{as} \quad N \to \infty,$$

we get $y = Tx$, concluding the proof. $\square$

Corollary 3.38 (Inverse mapping) *Let X, Y be Banach spaces over $\mathbb{K}$ and let $T \in \mathcal{L}(X, Y)$ injective and surjective. Then there exists $S \in \mathcal{L}(Y, X)$ inverse map of T, i.e.,*

$$T \circ S = Id \in \mathcal{L}(Y), \quad S \circ T = Id \in \mathcal{L}(X).$$

(Recall the notation $\mathcal{L}(X) = \mathcal{L}(X, X)$, and that Id is the identity map that associates to each point $x \in X$ the point x itself.)

Proof Let S be the inverse of T. Clearly S is a linear map from X *onto* Y (i.e., a surjective map from X to Y). Let U be an open subset of X, then the preimage $S^{-1}(U) = \{y \in Y : Sy = x \in U\} = T(U)$. $T(U)$ is open by the open mapping Theorem and S is hence continuous, thus, bounded from Y to X. $\square$

Definition 3.39 Let $(X, \|\cdot\|_X)$ and $(Y, \|\cdot\|_Y)$ be normed spaces over $\mathbb{K}$. For any pair $(x, y) \in X \times Y$ we set

$$\|(x, y)\|_{X \times Y} = \|x\|_X + \|y\|_Y, \qquad \text{the product norm on } X \times Y.$$

The space $(X \times Y, \|\cdot\|_{X \times Y})$ is a normed space, and if both X and Y are Banach spaces, then also $X \times Y$ is.

Definition 3.40 In the above notations, let $T : X \to Y$. We set

$$\Gamma(T) = \{(x, y) \in X \times Y : y = Tx\} = \{(x, Tx), x \in X\}.$$

We call $\Gamma(T)$ the graph of T, and in general we have $\Gamma(T) \subset X \times Y$, which is a linear subspace if the map T is linear. If T is linear, then the product norm in $X \times Y$ induces a norm on $\Gamma(T)$ given by

$$\|(x, y)\|_{\Gamma(T)} = \|(x, Tx)\|_{\Gamma(T)} = \|x\|_X + \|Tx\|_Y .$$

Theorem 3.41 *Let X and Y be Banach spaces over $\mathbb{K}$ and let $T : X \to Y$ be a linear map. Then $T \in \mathcal{L}(X, Y)$ if and only if $\Gamma(T)$ is closed in $X \times Y$.*

Proof We prove separately the two implications.

$\implies$ Suppose $T \in \mathcal{L}(X, Y)$ and take $\{(x_n, y_n)\}_n = \{(x_n, Tx_n)\}_n$ a sequence in $\Gamma(T)$, such that $(x_n, Tx_n) \to (x, y)$, as n tends to infinity, for some y in Y. This means that $\{x_n\}_n$ converges to x and that $\{Tx_n\}_n$ converges to y, but then by continuity of T we have that $y = Tx$ proving that the limit point (x, y) is in $\Gamma(T)$, thus, $\Gamma(T)$ is closed.

$\impliedby$ Viceversa suppose that $\Gamma(T)$ is a closed subspace of $X \times Y$ and introduce the projection operators of $X \times Y$:

$$\pi_1 : X \times Y \to X, \quad \pi_1(x, y) = x;$$
$$\pi_2 : X \times Y \to Y, \quad \pi_2(x, y) = y. \tag{3.16}$$

Clearly π_1 and π_2 are linear and continuous, and thus, restricted to $\Gamma(T)$ we have that $\pi_1|_{\Gamma(T)} \in \mathcal{L}(\Gamma(T), X)$ and $\pi_2|_{\Gamma(T)} \in \mathcal{L}(\Gamma(T), Y)$. Moreover, $\pi_1|_{\Gamma(T)}$ is trivially surjective, but also injective: indeed if $\pi_1(x, Tx) = \pi_1(z, Tz)$, then $x = z$ by definition of π_1. Thus, using inverse mapping Theorem there exists $\pi_1^{-1} \in \mathcal{L}(X, \Gamma(T))$ such that $\pi_1^{-1} : x \to (x, Tx)$. Consider now the composition $\pi_2 \circ \pi_1^{-1} : X \to Y$, it is a continuous linear operator from X to Y, and since $\pi_2 \circ \pi_1^{-1} = T$, we conclude. $\qquad\square$

3.5 Precompactness and Convexity

Proposition 3.42 *Let $(X, \|\cdot\|)$ be a normed space over $\mathbb{K}$ and consider a subset A of X. Then A is precompact (relatively compact) if and only if the following two conditions hold true:*

1. *A is bounded;*
2. *for any positive ε there exists a finite dimensional subspace F_ε of X such that*

$$\mathrm{dist}(x, F_\varepsilon) < \varepsilon, \quad \forall x \in A.$$

Proof We prove separately the two implications.

$\Longrightarrow$ Suppose that A is precompact, which means that for every fixed ε there exists a finite set of points $\{y_1, \ldots, y_{n_\varepsilon}\}$ in X such that A is contained in the union of balls of radius ε and center y_i. Thus, being contained in a finite union of bounded sets, A is clearly bounded. Next introduce the finite dimensional subspace

$$F_\varepsilon := \mathrm{lsp}(\{y_1, \ldots, y_{n_\varepsilon}\}).$$

We show that it satisfies condition 2. Indeed, for any point x in A, there exists an index k between 1 and n_ε such that $x \in B(y_k, \varepsilon)$. Thus, $\mathrm{dist}(x, F_\varepsilon) \le \|y_k - x\| < \varepsilon$.

$\Longleftarrow$ Suppose that conditions 1 and 2 hold and fix $\varepsilon > 0$. We construct a finite set of points $\{y_1, \ldots, y_{n_\varepsilon}\}$ such that $A \subseteq \bigcup_{i=1}^{n_\varepsilon} B(y_i, 2\varepsilon)$, concluding the proof. First we use boundedness of A to fix $M > 0$ such that $A \subseteq B(0, M)$. Given $\varepsilon > 0$ let F_ε be as in point 2. Consider now the set of points in F_ε having distance from A smaller then ε:

$$B = \{x \in F_\varepsilon : \mathrm{dist}(x, A) < \varepsilon\}.$$

Clearly B is bounded, and, being a subset of the finite dimensional set F_ε, it is precompact. Let hence $\{y_1, \ldots, y_{n_\varepsilon}\} \subseteq F_\varepsilon$ be such that

$$B \subseteq \bigcup_{i=1}^{n_\varepsilon} B(y_i, \varepsilon).$$

Since, by condition 2, for any x in A there exists $z \in F_\varepsilon$ such that $\|z - x\| < \varepsilon$, this implies that $z \in B$. Consequently there exists $1 \le k \le n_\varepsilon$ such that $\|z - y_k\| < \varepsilon$. Using triangular inequality we conclude that

$$\|x - y_k\| \le \|x - z\| + \|z - y_k\| < 2\varepsilon.$$

From the arbitrariness of x we conclude the proof since $A \subseteq \bigcup_{i=1}^{n_\varepsilon} B(y_i, 2\varepsilon)$. $\quad\square$

Proposition 3.43 *Let $(X, \|\cdot\|)$ be a normed space and $C \subseteq X$ a convex subset with nonempty interior. Let $f : C \to \mathbb{R}$ be such that*

$$f((1 - t)x + ty) \le (1 - t)f(x) + tf(y), \quad \forall x, y \in C \text{ and } \forall t \in [0, 1],$$

(i.e., f is convex). Given $x_0 \in \mathrm{int}(C)$ assume that there exists $\alpha \in \mathbb{R}$ and a positive value η such that

$$f(x) \le \alpha \quad \forall x \in C \cap B(x_0, \eta). \tag{3.17}$$

Then f is continuous at x_0.

Proof We may assume that $x_0 = 0$, $f(x_0) = 0$ and $\eta = 1$ (up to rescaling and translations). In this way hypothesis (3.17) of the proposition becomes

$$f(y) \leq \alpha, \quad \forall \, \|y\| < 1.$$

We now claim that for all x with norm smaller than ε we have $|f(x)| \leq \alpha\varepsilon$, concluding the proof. To prove this claim we start using convexity of f and obtain

$$f(x) = f\left(\varepsilon\frac{x}{\varepsilon} + (1 - \varepsilon)0\right) \leq \varepsilon f\left(\frac{x}{\varepsilon}\right).$$

If $\|x\| \leq \varepsilon$ we have $\left\|\frac{x}{\varepsilon}\right\| \leq 1$ and then $f(x) \leq \alpha\varepsilon$. In order to prove the opposite inequality we write

$$0 = f\left(\frac{\varepsilon}{1+\varepsilon}\left(-\frac{x}{\varepsilon}\right) + \frac{1}{1+\varepsilon}x\right) \leq \frac{\varepsilon}{1+\varepsilon}f\left(-\frac{x}{\varepsilon}\right) + \frac{1}{1+\varepsilon}f(x), \implies f(x) \geq -\varepsilon f\left(-\frac{x}{\varepsilon}\right).$$

If $\|x\| \leq \varepsilon$, then $\left\|-\frac{x}{\varepsilon}\right\| \leq 1$ obtaining that $f\left(-\frac{x}{\varepsilon}\right) \leq \alpha$. Hence $f(x) \geq -\alpha\varepsilon$, concluding the proof of the claim. $\qquad\square$

Definition 3.44 Let $(X, \|\cdot\|_X)$ and $(Y, \|\cdot\|_Y)$ be normed spaces and $A \subseteq X$ nonempty. We say that a function $f : A \to Y$ is Lipschitz continuous is there exists a positive constant L such that

$$\|f(x) - f(y)\|_Y \leq L \, \|x - y\|_X, \quad \forall x, y \in A.$$

Proposition 3.45 *Let* $(X, \|\cdot\|)$ *be a normed space and let* $A \subseteq X$ *nonempty. We define the function* $f : X \to \mathbb{R}$ *as*

$$f(x) := \text{dist}(x, A) = \inf\{\|x - y\|, \, y \in A\}.$$

The following hold:

1. f is Lipschitz continuous;
2. if A is convex, then f is convex.

Proof 1. From the definition of the function f we have that, given x in X

$$\text{dist}(x, A) \leq \|x - z\|, \quad \forall z \in A.$$

Thus, for every $\varepsilon > 0$ there exists a point x_ε in A such that

$$\|x - x_\varepsilon\| \leq \text{dist}(x, A) + \varepsilon.$$

Take any x, y in X, for any $z \in A$ we have

$$f(x) - f(y) \leq \|x - z\| - \|y - y_\varepsilon\| + \varepsilon \overset{z=y_\varepsilon}{\leq} \|x - y_\varepsilon\| - \|y - y_\varepsilon\| + \varepsilon \leq \|x - y\| + \varepsilon.$$

Analogously exchanging x with y we conclude that $|f(x) - f(y)| \leq \|x - y\| + \varepsilon$, for any positive ε.

2. Take $x, y \in X$ and $\varepsilon > 0$. Let x_ε and y_ε as in the proof of point 1. Then for any $t \in [0, 1]$ we have

$$
\begin{aligned}
f((1-t)x + ty) &= \mathrm{dist}((1-t)x + ty, A) \leq \|(1-t)x + ty - [(1-t)x_\varepsilon + ty_\varepsilon]\| \\
&= \|(1-t)(x - x_\varepsilon) + t(y - y_\varepsilon)\| \leq (1-t)\|x - x_\varepsilon\| + t\|y - y_\varepsilon\| \\
&\leq (1-t)(f(x) + \varepsilon) + t(f(y) + \varepsilon) = (1-t)f(x) + tf(y) + \varepsilon,
\end{aligned}
\tag{3.18}
$$

for any positive ε. $\qquad\square$

Definition 3.46 Let $(X, \|\cdot\|)$ be a normed space and $A \subseteq X$, $A \neq \emptyset$. Introduce the family

$$
\mathcal{U} := \{C \subseteq X, \text{ closed, convex } : A \subseteq C\}.
$$

We set $\overline{Co(A)} := \bigcap_{C \in \mathcal{U}} C$, and call this set the *closed convex hull of* A.

Exercise: if A is precompact, that $\overline{Co(A)}$ is precompact (Hint: use characterization of precompact sets given by Proposition 3.42).

Remark 3.47 In the above notation define

$$
C := \{z = \sum_{i=1}^{N_z} \lambda_i x_i, \ x_i \in A, \lambda_i > 0, \ \sum_{i=1}^{N_z} \lambda_i = 1\}.
$$

Then $\overline{Co(A)} = \overline{C}$.

Definition 3.48 Let $(X, \|\cdot\|)$ be a normed space. Given $A \subseteq X$ convex and nonempty, we say that a convex and nonempty subset $F \subseteq A$ is an (external) face of A if the following property holds: if there exists x, y in A and some $\lambda \in]0, 1[$ such that $x_0 := (1 - \lambda)x + \lambda y \in F$, then $x, y \in F$. In particular A is a face of A and every set containing a single point is a face of A.

A point $x_0 \in A$ is called an extremal point of A if for any $x, y \in A$ such that there exists $\lambda \in]0, 1[$ such that $x_0 = (1 - \lambda)x + \lambda y$, then $x = y$.

We denote by $\mathrm{extr}(A)$ the set of extremal points of A.

We conclude by stating the Krein-Milman Theorem, which is often used in applications.

Theorem 3.49 (Krein-Milman) *Let* $(X, \|\cdot\|)$ *be a normed space and let* $K \subseteq X$ *nonempty, convex and compact. Then*

1. $\mathrm{extr}(K) \neq \emptyset$;
2. $K = \overline{Co}(\mathrm{extr}(K))$.

3.6 Compact Operators

Definition 3.50 Let X, Y be Banach spaces and let $T \in \mathcal{L}(X, Y)$. We say that T is *compact* if the set $T(\overline{B}_1^X)$ is relatively compact in Y. Equivalently, T is compact if, given any $A \subseteq X$ bounded, $T(A)$ is relatively compact in Y.

Remark 3.51 If Y is finite dimensional, that any $T \in \mathcal{L}(X, Y)$ is compact.

Lemma 3.52 *Let X, Y, Z be Banach spaces and let $T \in \mathcal{L}(X, Y)$ and $S \in \mathcal{L}(Y, Z)$. If at least one among T and S is compact, then $S \circ T$ is compact too.*

Proof Let $A \subseteq X$ be a bounded set and suppose that T is compact. Let $(x_n)_n$ be a sequence in A, then $(Tx_n)_n$ is a sequence in $T(A)$ and since T is compact, it converges up to subsequence. Thus, there exists $(x_{n_k})_k$ such that $(Tx_{n_k})_k$ converges in $T(A)$. Now using the continuity of S, it follows that $(S(Tx_{n_k}))_k$ also converges, proving the compactness of $S \circ T$.

Now suppose that S is compact, since $T(A)$ is bounded, then $S(T(A))$ is relatively compact, thus $S \circ T$ is compact in this second case as well. $\qquad\square$

Lemma 3.53 *Let X, Y be Banach spaces and denote by $\mathcal{K} = \mathcal{K}(X, Y) \subset \mathcal{L}(X, Y)$ the set of compact operators. Then $\mathcal{K}$ is a closed subspace of $\mathcal{L}(X, Y)$.*

Proof It is straightforward that if $T, S \in \mathcal{K}$, then $T + S$ is an element of $\mathcal{K}$, since the sum of precompact sets is precompact, and also that for any $\alpha \in \mathbb{K}$, $\alpha T \in \mathcal{K}$. Thus, $\mathcal{K}$ is a subspace of $\mathcal{L}(X, Y)$. To prove closedness, let $(T_n)_n$ be a sequence in $\mathcal{K}$ converging to an element $T \in \mathcal{L}(X, Y)$. Fix $\varepsilon > 0$ and let $n \in \mathbb{N}$ such that $\|T_n - T\| \leq \varepsilon$. Since T_n is compact, we have

$$T_n(\overline{B}_1^X) \subseteq \bigcup_{\substack{j \in J \\ J \text{ finite}}} B^Y(T_n x_j, \varepsilon), \quad x_j \in \overline{B}_1^X, \ \forall j \in J.$$

Let $x \in \overline{B}_1^X$ arbitrary and $j \in J$:

$$\left\| T_n x - T_n x_j \right\|_Y < \varepsilon \implies \left\| Tx - Tx_j \right\|_Y \leq 3\varepsilon,$$

where for the last inequality we use triangular inequality, summing and subtracting both $T_n x$ and $T_n x_j$. $\qquad\square$

Chapter 4
Continuous Functions

In this section we study the properties of the space of continuous functions on a compact metric space using the tools that we previously introduced. Before starting, we preliminarily prove the following proposition, usually called *diagonal procedure*, that we will use many times in this section.

Proposition 4.1 *Let $\{(X_p, d_p),\ p \in \mathbb{N}\}$ be a countable family of metric spaces. For every $p \in \mathbb{N}$ let $(x_n^p)_n$ be a relatively compact sequence in (X_p, d_p). Then there exists a strictly increasing map $\phi : \mathbb{N} \to \mathbb{N}$ such that the sequence $(x_{\phi(n)}^p)_{n \in \mathbb{N}}$ is converging for every $p \in \mathbb{N}$.*

Proof Let $A_1 \subseteq \mathbb{N}$ be a countable family of indices such that $(x_n^1)_{n \in A_1}$ converges. By hypothesis the sequence $(x_n^2)_{n \in \mathbb{N}}$ is relatively compact. Thus, we can find $A_2 \subseteq A_1$ such that $(x_n^2)_{n \in A_2}$ converges. Obviously, $(x_n^1)_{n \in A_2}$ converges as well.

Iterating this procedure, define for any $p \in \mathbb{N}$ a countable set $A_p \subseteq A_{p-1}$ such that the sequence $(x_n^p)_{n \in A_p}$ converges. Defining $\phi(p) := (p+1)$-th element of A_p, we have by construction that $\phi(p) > \phi(p-1)$, for all $p \in \mathbb{N}$.

Additionally $\phi(n) \in A_n \subseteq A_p$, for $n \geq p$ and hence $(x_{\phi(n)}^p)_{n \geq p}$ converges for any $p \in \mathbb{N}$. Thus, $(x_{\phi(n)}^p)_{n \in \mathbb{N}}$ converges for any $p \in \mathbb{N}$. $\qquad\square$

Definition 4.2 Let (X, d) be a compact metric space (through the rest of this section we always assume that X is a space with at least two points). We introduce the spaces

$$\mathcal{C}(X) := \{u : X \to \mathbb{C},\ u \text{ continuous}\} \quad \text{and} \quad \mathcal{C}(X; \mathbb{R}) := \{u : X \to \mathbb{R},\ u \text{ continuous}\}. \quad (4.1)$$

1. We define the operations

$$(u + v)(x) := u(x) + v(x), \quad x \in X; \quad (4.2)$$

$$(\lambda u)(x) = \lambda u(x), \quad x \in X,\ \lambda \in \mathbb{K}, \quad (4.3)$$

© The Author(s), under exclusive license to Springer Nature Switzerland AG 2026
S. Zagatti, *Functional Analysis*, SISSA Springer Series 7,
https://doi.org/10.1007/978-3-032-24475-8_4

where u and v both lie in $C(X)$ or $C(X; \mathbb{R})$. The spaces $C(X)$ and $C(X; \mathbb{R})$, endowed with these two operations, are *linear spaces* over $\mathbb{C}$ and $\mathbb{R}$, respectively. The zero vector is clearly the function that associates $0 \in \mathbb{K}$ to every point x.

2. We set

$$(u \cdot v)(x) = u(x) \cdot v(x), \quad \forall x \in X; \tag{4.4}$$

$$\mathbb{1}(x) = 1 \in \mathbb{K}, \quad \forall x \in X. \tag{4.5}$$

The spaces $C(X)$ and $C(X; \mathbb{R})$ turn out to be commutative algebras with unity if endowed with this multiplicative operation.

We set some other notations that will be useful in this section:

$$u \vee v := \max\{u, v\}, \quad \text{i.e., } (u \vee v)(x) = \max\{u(x), v(x)\}, \quad \forall x \in X; \tag{4.6}$$

$$u \wedge v := \min\{u, v\} \quad \text{i.e., } (u \wedge v)(x) = \min\{u(x), v(x)\}, \quad \forall x \in X; \tag{4.7}$$

$$u^+ := u \vee 0 > 0 \text{ is called the } positive \ part \text{ of } u; \tag{4.8}$$

$$u^- = (-u) \vee 0 > 0 \text{ is called the } negative \ part \text{ of } u. \tag{4.9}$$

We observe that

$$u = u^+ - u^-; \quad |u| = u^+ + u^-; \quad u^- = (-u)^+;$$

$$u^+ = \frac{1}{2}(|u| + u); \quad u^- = \frac{1}{2}(|u| - u).$$

Hence $|u|$, u^+ and u^- are elements of $C(X; \mathbb{R})$. Consequently, also $u \vee v$ and $u \wedge v$ are elements of $C(X; \mathbb{R})$. Finally, we introduce the norm

$$\|u\|_{C(X)} := \max_{x \in X} |u(x)| = \|u\|_\infty, \tag{4.10}$$

generalizing (2.6) (and an analogous norm can be introduced in $C(X; \mathbb{R})$).

<u>Exercise:</u> prove that the spaces $(C(X), \|\cdot\|_\infty)$ and $(C(X; \mathbb{R}), \|\cdot\|_\infty)$ are Banach spaces.

4.1 Separability

The first important property that we study is separability. We prove that $C(X)$ and $C(X; \mathbb{R})$ are separable and we then identify subsets which are dense and countable in $C(X)$ and $C(X; \mathbb{R})$, respectively.

For this first result we work in $C(X)$ but the result in particular holds for $C(X; \mathbb{R})$.

Proposition 4.3 *The space $\mathcal{C}(X)$ is separable.*

Proof The main hypothesis to prove separability is the compactness of the space X, that we now use immediately: for every $n \in \mathbb{N}$ there exists a natural number $k_n \in \mathbb{N}$ and a set of points $\{x_i^n \in X, \ 1 \le i \le k_n\}$ such that

$$X \subseteq \bigcup_{1 \le i \le k_n} B\left(x_i^n, \frac{1}{n}\right). \tag{4.11}$$

Fix $n \in \mathbb{N}$ and, setting $J_n := \{1, \ldots, k_n\}$, for each $i \in J_n$ define the functions

$$X \ni x \mapsto \mathrm{d}(x, x_i^n) \in \mathbb{R}; \tag{4.12}$$

$$X \ni x \mapsto \left(\frac{1}{n} - \mathrm{d}(x, x_i^n)\right)^+ =: \tilde{\phi}_i^n(x), \tag{4.13}$$

where d is the distance on the space X. Clearly, $\tilde{\phi}_i^n$ is an element of $\mathcal{C}(X; \mathbb{R})$ and thus, of $\mathcal{C}(X)$. Define now

$$\phi_i^n(x) := \frac{\tilde{\phi}_i^n(x)}{\sum_{k \in J_n} \tilde{\phi}_k^n(x)}, \quad \forall x \in X, \ i \in J_n.$$

Thanks to (4.11) and (4.12), the function at the denominator is never vanishing, hence ϕ_i^n is well-defined and $\phi_i^n \in \mathcal{C}(X; \mathbb{R})$ for any $i \in J_n$. These other properties clearly hold:

1. $0 \le \phi_i^n \le 1, \quad \forall i \in J_n$;
2. $\phi_i^n(x) > 0, \iff \mathrm{d}(x, x_i^n) \le \frac{1}{n}$;
3. $\sum_{i \in J_n} \phi_i^n = 1, \quad$ on X.

We finally consider the subspaces of $\mathcal{C}(X)$ defined by

$$Y := \mathrm{lsp}_{\mathbb{C}}(\{\phi_i^n, \ i \in J_n, \ n \in \mathbb{N}\})$$

and

$$\tilde{Y} := \mathrm{lsp}_{\mathbb{Q}}(\{\phi_i^n, \ i \in J_n, \ n \in \mathbb{N}\})$$

(see Definition 2.14) and prove that Y is dense in $\mathcal{C}(X)$. Since $\tilde{Y}$ is countable and dense in Y, this concludes the proof. To this aim, for any $f \in \mathcal{C}(X)$ and $\varepsilon > 0$ we construct a function $g \in Y$ such that $\|g - f\|_\infty \le \varepsilon$. We use uniform continuity of f: there exists $\delta > 0$ such that

$$\mathrm{d}(x, y) \le \delta \implies |f(x) - f(y)| \le \varepsilon.$$

Take then $n \in \mathbb{N}$ such that $\frac{1}{n} < \delta$ and set

$$g(x) = \sum_{i \in J_n} f(x_i^n)\phi_i^n(x), \quad \forall x \in X.$$

Clearly $g \in Y$ and computing $g - f$ we see that, since $\phi_i^n(x)$ is nonzero if and only if $\mathrm{d}(x, x_i^n) < \frac{1}{n} < \delta$ and recalling that $\sum_{i \in J_n} \phi_i^n = 1$

$$g(x) - f(x) = \sum_{j \in J_n} f(x_j^n)\phi_j^n(x) - f(x) \sum_{i \in J_n} \phi_i^n(x) = \sum_{j \in J_n} (f(x_j^n) - f(x))\phi_j^n(x),$$

thus

$$|f(x_i^n) - f(x)| \le \varepsilon, \ \forall \mathrm{d}(x, x_i^n) < \frac{1}{n}, \ \implies \ |f(x) - g(x)| \le \varepsilon \ \ \forall x \in X,$$

i.e., $\|f - g\|_\infty \le \varepsilon$. $\qquad\qquad\qquad\qquad\qquad\qquad\qquad\qquad\qquad\qquad\qquad\qquad\square$

Lemma 4.4 (Dini) *Let $(f_n)_{n \in \mathbb{N}}$ be a sequence in $\mathcal{C}(X; \mathbb{R})$ and let $f \in \mathcal{C}(X; \mathbb{R})$. Assume that*

1. $f_n \le f_{n+1}, \quad \forall n \in \mathbb{N}$;
2. $f_n(x) \to f(x), \quad \forall x \in X$;

Then f_n converges uniformly to f, i.e., $\|f_n - f\|_\infty \to 0$.

Proof Fix $\varepsilon > 0$ and for any $n \in \mathbb{N}$ define the open set:

$$\Omega_n^\varepsilon := \{x \in X : f(x) - \varepsilon < f_n(x)\}.$$

Clearly $\Omega_n^\varepsilon \subseteq \Omega_{n+1}^\varepsilon$, for any $\varepsilon > 0$ and $n \in \mathbb{N}$. Moreover, using hypothesis 2 we see that given any x in X there exists $\bar{n} \in \mathbb{N}$ such that $x \in \Omega_{\bar{n}}^\varepsilon$. Hence, we have

$$X = \bigcup_{n=1}^{\infty} \Omega_n^\varepsilon$$

and using compactness of X we extract a finite subcovering that we denote by $\{\Omega_{n_1}^\varepsilon, \dots, \Omega_{n_k}^\varepsilon\}$. Hence

$$X = \bigcup_{i=1}^{k} \Omega_{n_i}^\varepsilon = \Omega_N^\varepsilon, \quad \text{with } N := \max\{n_1, \dots, n_k\}.$$

Thus, we have

$$f(x) - \varepsilon < f_n(x) \le f(x), \quad \forall x \in X, \quad \forall n \ge N.$$

By the arbitrariness of $\varepsilon > 0$ this means that f_n converges to f uniformly in $(\mathcal{C}(X), \|\cdot\|_\infty)$. $\qquad\qquad\qquad\qquad\qquad\qquad\qquad\qquad\qquad\qquad\qquad\square$

We now give a sufficient condition on the subsets of $\mathcal{C}(X; \mathbb{R})$ to ensure density $\mathcal{C}(X; \mathbb{R})$.

Definition 4.5 (*Lattice*) Take H a subset of $\mathcal{C}(X; \mathbb{R})$. We say that H is a lattice if

1. H is *separating* in $\mathcal{C}(X; \mathbb{R})$, i.e., for any pair of points x and y in X with $x \neq y$, there exists a function $h \in H$ such that

$$h(x) \neq h(y);$$

2. (lattice property) for all $u, v \in H$ we have

$$u \vee v \text{ and } u \wedge v \in H. \tag{4.14}$$

Proposition 4.6 *Let H be a* subspace *and a* lattice *of $\mathcal{C}(X; \mathbb{R})$ which contains the constant functions. Then H is dense in $\mathcal{C}(X; \mathbb{R})$.*

The proof of the proposition is a consequence of the following Lemma:

Lemma 4.7 *If $H \subseteq \mathcal{C}(X; \mathbb{R})$ satisfies lattice property* (4.14) *and if for all x_1, x_2 in X with $x_1 \neq x_2$, and for all $\alpha_1, \alpha_2 \in \mathbb{R}$ there exists $u \in H$ such that*

$$u(x_1) = \alpha_1 \quad \text{and} \quad u(x_2) = \alpha_2,$$

then H is dense in $\mathcal{C}(X; \mathbb{R})$.

Proof of Lemma 4.7. Let $f \in \mathcal{C}(X; \mathbb{R})$ and $\varepsilon > 0$. By the second hypothesis, for any $x \in X$ we have that for every $y \in X$ with $y \neq x$, there exists an element u_y of H such that

$$u_y(x) = f(x), \quad \text{and} \quad u_y(y) = f(y).$$

We now define the set

$$O_y := \{z \in X : u_y(z) > f(z) - \varepsilon\}.$$

Clearly x and y are both elements of O_y and we also observe that O_y is open. Thus,

$$X = \bigcup_{\substack{y \neq x \\ y \in X}} O_y, \quad \overset{\text{compactness}}{\Longrightarrow} \quad X = \bigcup_{i=1}^{r} O_{y_i},$$

for some $y_1, \ldots y_r \in X \setminus \{x\}$. We now set

$$v_x := \max\{u_{y_1}, \ldots, u_{y_r}\} \in H,$$

(recall (4.14)). It is immediate to check that $v_x(x) = f(x)$ and $v_x(z) > f(z) - \varepsilon$ for all $z \in X$. We now let x vary on X and consider the family of functions $\{v_x, \ x \in X\}$. Define the new family of sets

$$\tilde{O}_x = \{z \in X : \ v_x(z) < f(z) + \varepsilon\},$$

obtaining as above that

$$X = \bigcup_{x \in X} \tilde{O}_x, \quad \overset{\text{compactness}}{\Longrightarrow} \quad X = \bigcup_{i=1}^{p} \tilde{O}_{x_i}.$$

We define

$$v = \min\{v_{x_1}, \ldots, v_{x_p}\} \in H,$$

(recall (4.14)) and by construction we have

$$f - \varepsilon \leq v \leq f + \varepsilon,$$

concluding the proof. $\qquad\square$

Proof of Proposition 4.6. We apply Lemma 4.7: given $x_1 \neq x_2$ in X and α_1 and α_2 real numbers, take h in H such that $h(x_1) \neq h(x_2)$. Consider the following linear system in λ, μ:

$$\begin{cases} \lambda h(x_1) + \mu = \alpha_1 \\ \lambda h(x_2) + \mu = \alpha_2. \end{cases} \tag{4.15}$$

It has a unique solution $(\overline{\lambda}, \overline{\mu}) \in \mathbb{R}^2$, and we define

$$\overline{h}(x) = \overline{\lambda} h(x) + \overline{\mu}.$$

Clearly $\overline{h} \in H$ and by construction $\overline{h}$ satisfies the required condition $\overline{h}(x_1) = \alpha_1$ and $\overline{h}(x_2) = \alpha_2$. $\qquad\square$

Exercise: Let $X \subseteq \mathbb{R}^d$ be a compact subset, consider

$$H = \mathrm{Lip}(X; \mathbb{R}) = \{f \in \mathcal{C}(X; \mathbb{R}) : \exists L_f \geq 0, \ |f(x) - f(y)| \leq L_f |x - y|, \ \forall x, y \in X\}.$$

Prove that H is a subspace that satisfies the hypotheses of Proposition 4.6, thus, is dense in $\mathcal{C}(X; \mathbb{R})$.

Finally, we conclude the study of separability of the space of continuous functions proving that the space of real valued polynomials $p : X \to \mathbb{R}$ is dense in $\mathcal{C}(X; \mathbb{R})$. To this aim, we first prove the following lemma.

Lemma 4.8 *Let $H \subseteq C(X; \mathbb{R})$ be a subspace. Then H is a lattice if and only if*

$$f \in H \implies |f| \in H. \tag{4.16}$$

Proof We prove separately the two implications.

$\implies$ Suppose that H is a lattice. We observe that the zero function lies in H since H is a subspace of $C(X; \mathbb{R})$. Next, take $f \in H$ and write the absolute value of f as follows (recall (4.9)):

$$|f| = f^+ + f^- = f \vee 0 + (-f) \vee 0 \in H,$$

proving the first implication.

$\impliedby$ Viceversa take $u, v \in H$ and rewrite $u \vee v$ and $u \wedge v$ as:

$$u \vee v = (u - v)^+ + v = \frac{1}{2}(|u - v| + u - v) + v,$$

and analogously

$$u \wedge v = -(u - v)^- + v = -\frac{1}{2}(|u - v| - u + v) + v.$$

Using hypothesis (4.16) and the fact that H is a subspace, this proves that $u \vee v$ and $u \wedge v$ are elements of H, concluding the proof.

$\square$

Lemma 4.9 *There exists a sequence $(p_n)_{n \in \mathbb{N}}$ of real polynomials in one variable that uniformly converges to the absolute value function $x \mapsto |x|$ in the interval $[-1, 1]$, i.e.,*

$$\max_{x \in [-1,1]} |p_n(x) - |x|| \to 0, \quad \text{as } n \to \infty.$$

Proof We construct $(p_n)_{n \in \mathbb{N}}$ by induction. Set $p_0 = 0$ and define

$$p_{n+1}(x) = p_n(x) + \frac{1}{2}(x^2 - p_n(x)^2), \quad \forall n \in \mathbb{N}. \tag{4.17}$$

We now show that

$$0 \le p_n(x) \le p_{n+1}(x) \le |x|, \quad \forall x \in [-1, 1], \ \forall n \in \mathbb{N}. \tag{4.18}$$

If this chain of inequalities holds, then for every fixed $x \in [-1, 1]$, the real sequence $(p_n(x))_{n \in \mathbb{N}}$ is bounded and increasing, hence there exists $f : [-1, 1] \to \mathbb{R}$ such that $p_n(x) \to f(x)$ for all $x \in [-1, 1]$. To identify the function f we pass to the limit in the recursion formula (4.17), obtaining

$$f(x) = f(x) + \frac{1}{2}(x^2 - f(x)^2) \implies f(x) = |x|, \quad \forall x \in [-1, 1].$$

This concludes the proof, since Dini's Lemma 4.4 implies that $(p_n)_{n \in \mathbb{N}}$ uniformly converges to the absolute value function in $[-1, 1]$.

Hence, we are left to prove the inequalities in (4.18). We proceed inductively: the case $n = 0$ is trivially true, since (4.18) reads

$$0 \le 0 \le \frac{x^2}{2} \le |x|.$$

We now assume that (4.18) holds, and we prove that it hold for $n + 1$: first we recall that, by definition

$$p_{n+2}(x) = p_{n+1}(x) + \frac{1}{2}(x^2 - p_{n+1}(x)^2)$$

so that, from the inductive hypothesis $p_{n+1}(x) \le |x|$, we deduce that $(x^2 - p_{n+1}(x)^2) \ge 0$ which in turn implies $p_{n+1} \le p_{n+2}$.

Finally, to prove the third inequality in (4.18) we write:

$$\begin{aligned}
p_{n+2}(x) &= p_{n+1}(x) + \frac{1}{2}(x^2 - p_{n+1}(x)^2) \\
&= |x| - (|x| - p_{n+1}(x)) + \frac{1}{2}(|x| - p_{n+1}(x))(|x| + p_{n+1}(x)) \qquad (4.19) \\
&= |x| - \underbrace{(|x| - p_{n+1}(x))}_{\ge 0}[1 - \frac{1}{2}\underbrace{(|x| + p_{n+1}(x))}_{\le 2|x|}].
\end{aligned}$$

Again by inductive hypothesis, $p_{n+1}(x) \le |x|$, thus, substituting in previous equation we obtain $p_{n+2}(x) \le |x|$, for all $x \in [-1, 1]$. $\qquad \square$

Definition 4.10 Let $H \subseteq \mathcal{C}(X; \mathbb{R})$ be a subspace. We say that H is a subalgebra of $\mathcal{C}(X; \mathbb{R})$ if for all u, v in H we have $uv \in H$. Equivalently, since

$$uv = \frac{1}{2}((u + v)^2 - u^2 - v^2),$$

H is a subalgebra if for any $f \in H$ also $f^2 \in H$.

Proposition 4.11 *Let $H \subseteq \mathcal{C}(X; \mathbb{R})$ be a separating subalgebra containing constant functions. Then H is dense in $\mathcal{C}(X; \mathbb{R})$.*

Proof First we remark that, if $(f_n)_n$ and $(g_n)_n$ are sequences in H that uniformly converge to f and g respectively, then $(f_n g_n)_n$ uniformly converges to fg. Thus, the closure $\overline{H}$ is a separating subalgebra containing constant functions. We now show that $\overline{H}$ is a lattice, proving that it is dense in $\mathcal{C}(X; \mathbb{R})$ and then, consequently, that

H is dense in $\mathcal{C}(X; \mathbb{R})$. To this aim, we consider $(p_n)_n$ the sequence of the previous Lemma, and, given $f \in \mathcal{C}(X; \mathbb{R})$, with $f \neq 0$, we have that

$$p_n\left(\frac{f}{\|f\|_\infty}\right) \to \frac{|f|}{\|f\|_\infty}, \quad \text{uniformly.}$$

Thus, recalling Proposition 4.6, we have

$$|f| = \lim_{n \to \infty} \|f\|_\infty \, p_n\left(\frac{f}{\|f\|_\infty}\right), \quad \text{in } \mathcal{C}(X; \mathbb{R}). \tag{4.20}$$

Observe now that since H is an algebra and f is an element of H, also $p_n(f)$ is; moreover, since $f \in \overline{H}$, then also $p_n(f) \in \overline{H}$. Hence (4.20) implies that if $f \in \overline{H}$ also $|f| \in \overline{H}$, proving that $\overline{H}$ is a lattice. $\qquad\square$

Corollary 4.12 (Stone-Weierstrass Theorem) *Let $X \subseteq \mathbb{R}^d$ be compact and consider the family*

$$\mathcal{P} := \{p : X \to \mathbb{R}, \quad \text{real polynomial in } d \text{ variables}\}.$$

Then $\mathcal{P}$ is dense in $\mathcal{C}(X; \mathbb{R})$.

Indeed, $\mathcal{P}$ is a separating subalgebra containing the constant functions.

We conclude this section stating the analogous result for the space $\mathcal{C}(X)$ of complex-valued continuous functions over the compact set X.

Corollary 4.13 *Let $H \subseteq \mathcal{C}(X)$ be a separating subalgebra containing constant functions. Assume in addition that H is* selfconjugate, *i.e., $f \in H$ implies $\overline{f} \in H$ (where $\overline{z}$ denotes the complex conjugate of a complex number z). Then H is dense in $\mathcal{C}(X)$.*

Proof Let

$$H_\mathbb{R} := \{f \in H : f(x) \in \mathbb{R} \; \forall x \in X\}.$$

For every $f \in H$ we have that

$$\mathrm{Re}(f) = \frac{f + \overline{f}}{2} \in H_\mathbb{R} \quad \text{and} \quad \mathrm{Im}(f) = \frac{f - \overline{f}}{2i} \in H_\mathbb{R}.$$

$H_\mathbb{R}$ is a separating subalgebra of $\mathcal{C}(X; \mathbb{R})$ containing constant functions, hence it is dense in $\mathcal{C}(X; \mathbb{R})$. In addition

$$\mathcal{C}(X) = \mathcal{C}(X; \mathbb{R}) + i\mathcal{C}(X; \mathbb{R})$$

and $H = H_\mathbb{R} + i H_\mathbb{R}$. $\qquad\square$

4.2　Compactness and the Ascoli-Arzelà Theorem

We now turn to the study of compact subsets of the spaces of continuous functions. In particular, we prove the Ascoli-Arzelà Theorem, that gives necessary and sufficient conditions for compactness.

Definition 4.14 (*Equicontinuity*) Let $H \subseteq \mathcal{C}(X)$. H is equicontinuous if

$$\forall \varepsilon > 0 \; \exists \, \delta > 0 \text{ such that } \forall x, y \in X \text{ with } \mathsf{d}(x, y) \leq \delta, \text{ we have } |f(x) - f(y)| \leq \varepsilon, \; \forall f \in H.$$

Let us specify further the definition of equicontinuity in the particular case of a single point:

Definition 4.15 Let $H \subseteq \mathcal{C}(X)$ and let $x_0 \in X$. Then H is equicontinuous at the point x_0 if

$$\forall \varepsilon > 0 \; \exists \, \delta > 0 \text{ such that } \forall x \in X \text{ with } \mathsf{d}(x, x_0) \leq \delta, \text{ we have } |f(x) - f(x_0)| \leq \varepsilon, \; \forall f \in H.$$

Clearly if H is equicontinuous on X, it is equicontinuous at any point of X, thus, this last definitions allows to state the following:

Criterion for non-equicontinuity of H: if there exists $x_0 \in X$, a sequence $(f_n)_n \subseteq H$ and a sequence $(x_n)_n \subseteq X$, with $x_n \to x_0$ and such that $|f_n(x_n) - f_n(x_0)| \geq \gamma > 0$ for all $n \in \mathbb{N}$ and for some $\gamma > 0$, then H is not equicontinuous on X. This criterion is very useful in applications and exercises.

We now give some examples of equicontinuous and nonequicontinuous families:

1. Singleton: let X a compact metric space, $f \in \mathcal{C}(X)$ and $H = \{f\}$. Since f is uniformly continuous (by the Heine-Cantor Theorem), then H is equicontinuous.
2. Finite family: let X be a compact metric space and $f_1, \ldots, f_n \in \mathcal{C}(X)$. The family $H = \{f_1, \ldots, f_n\}$ is equicontinuous thanks to the uniform continuity of all the f_i, for $i = 1, \ldots, n$.
3. In general, a *finite union of equicontinuous families* is equicontinuous.
4. Equilipschitz family: let X be a compact metric space, $L > 0$ and let $H = \{u : X \to \mathbb{R} \; : \; |u(x) - u(y)| \leq L\mathsf{d}(x, y)\}$. Then H is equicontinuous. Indeed, let $\varepsilon > 0$ and take $\delta = \frac{\varepsilon}{L}$. If $\mathsf{d}(x, y) \leq \delta = \frac{\varepsilon}{L}$ then $|u(x) - u(y)| \leq \varepsilon$ for all $u \in H$.
5. Converging sequence: let X be a compact metric space and take a sequence $(f_n)_n \in \mathcal{C}(X)$, with $f_n \to f$ in $\mathcal{C}(X)$. Then $(f_n)_n$ is equicontinuous. Indeed, f is uniformly continuous, so that for any fixed $\varepsilon > 0$ there exists a positive δ such that $|f(x) - f(y)| \leq \varepsilon$ for all $x, y \in X$ with $\mathsf{d}(x, y) \leq \delta$. Next, choose $N_\varepsilon \in \mathbb{N}$ such that $\|f_n - f\|_\infty \leq \varepsilon$ for all $n \geq N_\varepsilon$. Then we have, for every $x, y \in X$, with $\mathsf{d}(x, y) \leq \delta$:

$$|f_n(x) - f_n(y)| \leq |f_n(x) - f(x)| + |f(x) - f(y)| + |f(y) - f_n(y)| \leq 3\varepsilon.$$

Hence, $H_1 := \{f_n, \ n \geq N_\varepsilon\}$ is equicontinuous. Consider now the family $H_2 := \{f_1, \ldots, f_{N_\varepsilon - 1}\}$. Since it is a finite family of functions, it is equicontinuous. Therefore $H = H_1 \cup H_2$ is equicontinuous being a finite union of equicontinuous families.

6. Let $X = [0, 1] \subseteq \mathbb{R}$ and consider $H = \{u_\alpha, \alpha \geq 1\}$, where $u_\alpha(t) = t^\alpha$ for $t \in [0, 1]$. Using the previously stated criterion we now show that H is not equicontinuous: we set $t_0 = 1$ and we look for a sequence $(t_n)_n \subseteq [0, 1]$ such that $t_n \to 1^-$ and $|u_n(t) - u_n(1)| = 1 - t^n \geq \beta$ for some β. Setting $t_n = \frac{1}{2^n}$, we obtain $|u_n(t) - u_n(1)| \geq 4 - 2^{-n} > \frac{1}{2} > 0$. that the family $H_\varepsilon = \{u_\alpha : [0, 1 - \varepsilon] \to \mathbb{R}\}_{\alpha \geq 1}$ with u_α defined as above, is an equicontinuous family for every $\varepsilon \in (0, 1)$.

7. Let $X = [0, 2\pi]$ and $H = \{u_\alpha : X \to \mathbb{R}, \ u_\alpha(t) = \sin \alpha t, \ \alpha \geq 1\}$. We check that H is not equicontinuous at any point of X. Take indeed $t \in X$ and choose for example $\beta = 1, t_n = t + \frac{\pi}{2n} \to t$, as $n \to \infty$. Then $|\sin(nt_n) - \sin(nt)| \geq \beta > 0$ for any $n \in \mathbb{N}$. Thus, H is not equicontinuous at any point $t \in [0, 2\pi]$.

Lemma 4.16 *Let $(f_n)_n$ be an equicontinuous sequence in $\mathcal{C}(X)$. Assume that there exists $\mathcal{D} \subset X$ dense in X and such that the sequence $(f_n(y))_n$ converges for any y in $\mathcal{D}$. Then there exists $f \in \mathcal{C}(X)$ such that f_n converges to f uniformly.*

Proof We show that $(f_n)_n$ is a Cauchy sequence in $\mathcal{C}(X)$. To this aim, we fix $\varepsilon > 0$ and let $\delta > 0$ be such that $|f_n(x) - f_n(y)| \leq \varepsilon$ for all pairs of points x, y such that $\mathrm{d}(x, y) \leq \delta$ and for all n. We use now compactness of X to find $\{\tilde{y}_j, \ j \in J \subseteq \mathbb{N}, |J| < \infty\}$ a finite collection of points such that

$$X \subseteq \bigcup_{j \in J} B\left(\tilde{y}_j, \frac{\delta}{2}\right).$$

Using the density of the set $\mathcal{D}$ in X, for every $j \in J$ we select y_j a point in $\mathcal{D} \cap B(\tilde{y}_j, \frac{\delta}{2})$. We can now prove that $(f_n)_n$ is a Cauchy sequence in $\mathcal{C}(X)$. Fix $x \in X$ and let $n, m \in \mathbb{N}$. Take $y_j \in \mathcal{D}$ such that $\mathrm{d}(x, y_j) < \delta$. Then

$$|f_n(x) - f_m(x)| \leq |f_n(x) - f_n(y_j)| + |f_n(y_j) - f_m(y_j)| + |f_m(y_j) - f_m(x)|$$
$$\leq 2\varepsilon + |f_n(y_j) - f_m(y_j)|.$$

$$(4.21)$$

Since J is finite there exists $N \in \mathbb{N}$ such that, for all $n, m \geq N$ we have $|f_n(y_j) - f_m(y_j)| \leq \varepsilon$, for all $j \in J$. Hence $\|f_n - f_m\|_\infty \leq 3\varepsilon$. $\qquad \square$

Theorem 4.17 (Ascoli-Arzelà) *Let $H \subseteq \mathcal{C}(X)$. Then H is relatively compact if and only if H is bounded and equicontinuous.*

Proof We prove separately the two implications.

$\implies$ Suppose that H is precompact. Boundedness follows automatically, thus, we are left with proving equicontinuity. To this aim, we fix $\varepsilon > 0$ and take $f_1, \ldots, f_N \in \mathcal{C}(X)$ (with $N = N_\varepsilon$) such that

$$H \subseteq \bigcup_{j=1}^{N} B_{\mathcal{C}(X)}(f_j, \varepsilon).$$

Since $\{f_1, \ldots, f_N\}$ is a finite set of functions, it is equicontinuous. Hence, there exists a positive δ such that $|f_j(x) - f_j(y)| \le \varepsilon$ for all x, y with $\mathrm{d}(x, y) \le \delta$ and for all $j = 1, \ldots, N$. $f \in H$ and take $k \in \{1, \ldots, N\}$ such that $\|f - f_k\|_\infty \le \varepsilon$. Then we have

$$|f(x) - f(y)| \le |f(x) - f_k(x)| + |f_k(x) - f_k(y)| + |f_k(y) - f(y)| \le 2\varepsilon + |f_k(x) - f_k(y)|.$$

If we impose $\mathrm{d}(x, y) \le \delta$ we obtain $|f(x) - f(y)| \le 3\varepsilon$.

$\Longleftarrow$ Assume now that H is bounded and equicontinuous. In order to prove precompactness we consider $(f_n)_n$ a sequence in H, and we claim that there exists a subsequence $(f_{n_k})_k$ and a function $f \in \mathcal{C}(X)$ such that $(f_{n_k})_k$ converges to f uniformly. First, we note that since X is compact, the space $\mathcal{C}(X)$ is separable (see 4.3), i.e., there exists $\mathcal{D} \subseteq X$ countable and dense in X. We now apply the diagonal procedure 4.1 to the countable family of sequences $(f_n(y))_n$, $y \in \mathcal{D}$. The boundedness of H implies that each of these sequences is bounded in $\mathbb{C}$, thus, $(f_n(y))_n$ is relatively compact. Using diagonal procedure 4.1 there exists a function $\phi : \mathbb{N} \to \mathbb{N}$ strictly increasing, such that $(f_{\phi(k)}(y))_{k \in \mathbb{N}}$ converges for all $y \in \mathcal{D}$. Hence, from previous Lemma 4.16 we conclude that there exits $f \in \mathcal{C}(X)$ such that $f_{n_k} \to f$ uniformly, for $n_k = \phi(k)$.

$\square$

Chapter 5
Hilbert Spaces

We now turn to the study Hilbert spaces, a subclass of Banach spaces, characterized by the possibility to define an inner product on them. We will explain how this allows for the generalization of geometric notions such as angles, distances, orthogonality and projection to infinite-dimensional spaces. The most familiar examples include Euclidean spaces, but examples of Hilbert spaces also include functions and sequence spaces, making them essential for applications.

5.1 Pre-Hilbert and Hilbert Spaces

Definition 5.1 (*Pre-Hilbert spaces*) Let X be a linear space over $\mathbb{K}$ and let $(\cdot, \cdot) : X \times X \to \mathbb{K}$ be a map satisfying the following properties:

1. $(x, x) \geq 0$, for all $x \in X$, and $(x, x) = 0$ if and only if $x = 0$;
2. $(x + y, z) = (x, z) + (y, z)$ for all $x, y, z \in X$;
3. $(\alpha x, y) = \alpha(x, y)$ for all $x, y \in X$ and $\alpha \in \mathbb{K}$;
4. $(x, y) = \overline{(y, x)}$ for all $x, y \in X$ (here $\bar{\ }$ denotes the complex conjugate);
4'. if X is a real vector space $(x, y) = (y, x)$ for all $x, y \in X$.

$(\cdot, \cdot)$ is called inner or scalar product and the pair $(X, (\cdot, \cdot))$ is said to be an inner product space, or pre-Hilbert space.

Remark that Properties 3. and 4. in previous definition imply that $(x, \alpha y) = \overline{\alpha}(x, y)$ for all $x, y \in X$ and for all $\alpha \in \mathbb{C}$.

Notation

– Let $x, y \in X$ be such that $(x, y) = 0$. We say that x and y are *orthogonal* and we write $x \perp y$;

© The Author(s), under exclusive license to Springer Nature Switzerland AG 2026
S. Zagatti, *Functional Analysis*, SISSA Springer Series 7,
https://doi.org/10.1007/978-3-032-24475-8_5

- given $E, F \subseteq X$ such that $(x, y) = 0$ for all $x \in E$ and $y \in F$, we say that E and F are orthogonal and write $E \perp F$;
- we set
$$\|x\| := (x, x)^{\frac{1}{2}}, \quad \text{for all } x \in X, \tag{5.1}$$

and we will show that it is a norm;
- we say that a set $S := \{x_\alpha, \ \alpha \in \mathcal{A}\} \subseteq X$ is an *orthogonal* system if $(x_\alpha, x_\beta) = 0$ for all $\alpha, \beta \in \mathcal{A}$, with $\alpha \neq \beta$;
- we say that a set $S := \{x_\alpha, \ \alpha \in \mathcal{A}\} \subseteq X$ is an *orthonormal* system if $(x_\alpha, x_\beta) = 0$ for all $\alpha, \beta \in \mathcal{A}$, $\alpha \neq \beta$ and $(x_\alpha, x_\alpha) = 1$, for all $\alpha \in \mathcal{A}$;

Remark 5.2 From the Definition 5.1 of inner product, we deduce immediately the following properties:

1. $(0, x) = (x, 0) = 0$ for all $x \in X$; indeed, $(0, x) = (0y, x) = 0(y, x) = 0$;
2. for any $x, y \in X$, using points 2 and 4 of Definition 5.1 we have

$$\|x + y\|^2 = (x + y, x + y) = (x, x) + (x, y) + (y, x) + (y, y) =$$
$$\|x\|^2 + (x, y) + \overline{(x, y)} + \|y\|^2 =$$
$$\|x\|^2 + \|y\|^2 + 2\operatorname{Re}(x, y);$$

and if $x \perp y$, then $\|x + y\|^2 = \|x\|^2 + \|y\|^2$;
3. arguing as in the previous point, if $S = \{x_i, \ i \in I, \ I \text{ finite }\}$ is an orthogonal system, then

$$\left\| \sum_{i \in I} x_i \right\|^2 = \sum_{i \in I} \|x_i\|^2 .$$

Lemma 5.3 (Pythagorean Theorem) *Let* $S = \{x_i, \ i \in I, \ I \text{ finite }\}$ *be an orthonormal system in* X. *Then, for every* $x \in X$ *we have*

$$\|x\|^2 = \sum_{i \in I} |(x, x_i)|^2 + \left\| x - \sum_{i \in I} (x, x_i) x_i \right\|^2 .$$

Proof We set

$$y = x - \sum_{i \in I} (x, x_i) x_i \quad \text{and} \quad z = \sum_{j \in I} (x, x_j) x_j.$$

From point 3 in Remark (5.2) we see that $\|z\|^2 = \sum_{j \in I} |(x, x_j)|^2$, thus, we conclude the proof if we show that $y \perp z$. To prove this we compute

$$\left(x - \sum_{i \in I} (x, x_i) x_i, \sum_{j \in I} (x, x_j) x_j \right) = \sum_{j \in I} (x, x_j)\overline{(x, x_j)} - \sum_{i \in I} \sum_{j \in I} (x, x_i)\overline{(x, x_j)}(x_i, x_j)$$
$$= \sum_{j \in I} |(x, x_j)|^2 - \sum_{i \in I} \sum_{j \in I} (x, x_i)\overline{(x, x_j)}\delta_{ij} = 0,$$

where $\delta_{ij} = 0$ if $i \neq j$ and $\delta_{ij} = 1$ if $i = j$. $\qquad\square$

Lemma 5.4 (Bessel inequality) *Let $S = \{x_i, i \in I \subseteq \mathbb{N}\}$ be an orthonormal system in X. Then for every $x \in X$ we have*

$$\|x\|^2 \geq \sum_{i \in I} |(x, x_i)|^2.$$

Proof Take any finite subset J of I. Using the Pythagorean Theorem 5.3 we have

$$\|x\|^2 \geq \sum_{i \in J} |(x, x_i)|^2.$$

Hence

$$\|x\|^2 \geq \sup_{\substack{J \in I \\ J\,finite}} \sum_{i \in J} |(x, x_i)|^2 = \sum_{i \in I} |(x, x_i)|^2,$$

concluding the proof. $\qquad\square$

Corollary 5.5 (Schwarz inequality) *For every $x, y \in X$ we have*

$$|(x, y)| \leq \|x\|\,\|y\|. \tag{5.2}$$

Proof If $y = 0$ then the claim is trivially true. Thus, suppose that $y \neq 0$ and consider the orthonormal system $S := \{\frac{y}{\|y\|}\}$. Using Bessel inequality:

$$\|x\|^2 \geq \left|\left(x, \frac{y}{\|y\|}\right)\right|^2 = \frac{|(x, y)|^2}{\|y\|^2}.$$

Hence $|(x, y)| \leq \|x\|\,\|y\|$. $\qquad\square$

Corollary 5.6 *For every $x, y \in X$ we have $\|x + y\| \leq \|x\| + \|y\|$.*

Proof Using point 2 of Remark 5.2 and Schwarz inequality (5.2) we have

$$\begin{aligned}
\|x + y\|^2 &= \|x\|^2 + \|y\|^2 + 2\operatorname{Re}(x, y) \leq \|x\|^2 + \|y\|^2 + 2|(x, y)| \\
&\leq \|x\|^2 + \|y\|^2 + 2\|x\|\,\|y\| = (\|x\| + \|y\|)^2,
\end{aligned} \tag{5.3}$$

concluding the proof. $\qquad\square$

Proposition 5.7 $(X, \|\cdot\|)$ *is a normed space.*

Proof From previous discussion the properties of the norm follow:

- $\|x\| = (x, x)^{\frac{1}{2}} \geq 0$, for all $x \in X$, and $\|x\| = 0$ if and only if $x = 0$;
- $\|\lambda x\|^2 = (\lambda x, \lambda x)^2 = \lambda\overline{\lambda}(x, x) = |\lambda|^2 \|x\|^2$, for all $\lambda \in \mathbb{C}$ and $x \in X$;
- from Corollary 5.6 it follows that $\|x + y\| \leq \|x\| + \|y\|$, for all $x, y \in X$.

$\square$

We also remark that the continuity of the map $X \times X \ni \{x, y\} \mapsto (x, y) \in \mathbb{K}$ follows from Schwarz inequality.

Remark 5.8 (a) We can rewrite

$$\|x\| = \sup_{\|y\|=1} |(x, y)|.$$

Indeed, from Schwarz inequality we have that

$$|(x, y)| \leq \|x\| \quad \forall \|y\| = 1,$$

and choosing $y = \frac{x}{\|x\|}$ we have

$$(x, y) = \|x\|.$$

(b) *Parallelogram identity:* given $x, y \in X$ we have

$$\|x + y\|^2 + \|x - y\|^2 = 2(\|x\|^2 + \|y\|^2). \tag{5.4}$$

Indeed,

$$\|x + y\|^2 + \|x - y\|^2 = \|x\|^2 + \|y\|^2 + 2\operatorname{Re}(x, y) + \|x\|^2 + \|y\|^2 - 2\operatorname{Re}(x, y).$$

Definition 5.9 (*Hilbert space*) A pre-Hilbert space that, as a normed space, turns out to be complete (i.e., a *Banach* space) is called a Hilbert space.

Theorem 5.10 (Completion) *Let* $(X, (\cdot, \cdot))$ *be a pre-Hilbert space. Then there exists a Hilbert space* $(\tilde{X}, \langle \cdot, \cdot \rangle)$ *and a map* $h : X \to \tilde{X}$ *such that*

1. $h(X)$ *is dense in* $\tilde{X}$;
2. $\langle h(x), h(y) \rangle = (x, y)$ *for all* $x, y \in X$.

The first examples to have in mind of finite dimensional Hilbert spaces are $\mathbb{R}^d$, equipped with the following scalar product:

$$(\cdot,\cdot): \mathbb{R}^d \times \mathbb{R}^d \to \mathbb{R}, \quad (x,y) = \sum_{i=1}^{d} x_i y_i, \quad \text{where } x = (x_1,\ldots,x_d),\, x_i \in \mathbb{R},$$

$$y = (y_1,\ldots,y_d),\, y_i \in \mathbb{R},$$

$$(5.5)$$

and $\mathbb{C}^d$, where the scalar product is analogously defined as

$$(\cdot,\cdot): \mathbb{C}^d \times \mathbb{C}^d \to \mathbb{C}, \quad (x,y) = \sum_{i=1}^{d} x_i \overline{y_i}, \quad \text{where } x = (x_1,\ldots,x_d),\, x_i \in \mathbb{C},$$

$$y = (y_1,\ldots,y_d),\, y_i \in \mathbb{C}.$$

$$(5.6)$$

5.2 Orthogonality and Projection Map

We now introduce projections of points and projector maps, which are concepts that arise from the definition of orthogonality and are peculiar of Hilbert spaces. From now on, we denote by $\mathcal{H}$ a generic Hilbert space.

Proposition 5.11 *Let $\mathcal{H}$ be a Hilbert space and let $C \subseteq \mathcal{H}$ be convex, closed and nonempty. Then for all $x \in \mathcal{H}$ there exists a unique point $P_C x$ in C such that*

$$\|x - P_C x\| = \mathrm{dist}(x, C) = \inf\{\|x - y\| : y \in C\}.$$

$P_C x$ is called projection *of x on C.*

Proof Fix $x \in \mathcal{H}$ and let $(y_n)_n$ be a sequence in C such that $\|x - y_n\| \to \mathrm{dist}(x, C)$ $=: \mathrm{d}$ (the existence of such sequence $(y_n)_n$ follows from the definition of inf). We now show that $(y_n)_n$ is a Cauchy sequence. Using the parallelogram identity (5.4) on the vectors $z_1 := x - y_n$ and $z_2 = x - y_m$, $n, m \in \mathbb{N}$ we have

$$\|(x - y_n) + (x - y_m)\|^2 + \|(x - y_n) - (x - y_m)\|^2 = 2(\|x - y_n\|^2 + \|x - y_m\|^2).$$

Hence,

$$\|y_n - y_m\|^2 + \|2x - (y_n + y_m)\|^2 = 2(\|x - y_n\|^2 + \|x - y_m\|^2),$$

but

$$\|2x - (y_n + y_m)\|^2 = 4\left\|x - \underbrace{\frac{y_n + y_m}{2}}_{\in C}\right\|^2 \geq 4\mathrm{d}^2$$

implying that

$$\|y_n - y_m\|^2 \le 2(\underbrace{\|x - y_n\|^2}_{\to d^2} + \underbrace{\|x - y_m\|^2}_{\to d^2}) - 4d^2 \to 0, \quad \text{as } n, m \to +\infty,$$

proving that $(y_n)_n$ is a Cauchy sequence and it thus converges to a point y, which, from closedness of C belongs to C. Finally, using continuity of the norm we have that $\|x - y\| = d$, thus, we set $P_C x := y$.

We conclude the proof showing uniqueness of $P_C y$: suppose that there exist two points $y_1, y_2 \in C$ such that

$$\|x - y_1\| = \|x - y_2\| = d = \text{dist}(x, C).$$

Rewriting again the parallelogram identity (5.4) with $z_1 := x - y_1$ and $z_2 := x - y_2$ we obtain

$$\|y_1 - y_2\|^2 = 2(d^2 + d^2) - 4\Big\|\underbrace{x - \frac{y_1 + y_2}{2}}_{\in C}\Big\|^2 \le 4d^2 - 4d^2 = 0,$$

hence $y_1 = y_2$. $\qquad\square$

Remark 5.12 If C is a closed subspace of $\mathcal{H}$ we define te map

$$H \ni x \mapsto P_C x \in C$$

and call it *orthogonal projector on* C. We will prove that $P_C \in \mathcal{L}(\mathcal{H})$.

Definition 5.13 Let $\mathcal{H}$ be a Hilbert space and let $S \subseteq \mathcal{H}$, $S \ne \emptyset$. The *orthogonal complement of* S is the set

$$S^\perp := \{x \in \mathcal{H} : (x, y) = 0, \ \forall y \in S\}.$$

Lemma 5.14 *1. $S^\perp$ is a closed subspace of $\mathcal{H}$;*
  2. $(\mathrm{lsp}(S))^\perp = S^\perp$;
  $$3. \ \ S \cap S^\perp = \begin{cases} \{0\} & \textit{if } 0 \in S \\ \emptyset & \textit{if } 0 \notin S. \end{cases}$$

Proof 1. To prove closedness of the set $S^\perp$ let $(x_n)_n$ be a sequence in $S^\perp$, converging to some $x \in \mathcal{H}$. Then for every $y \in S$ we have $0 = (x_n, y) \to (x, y) = 0$ by continuity of the scalar product. Thus, $x \in S^\perp$.
  2. Clearly $S \subseteq \mathrm{lsp}(S)$, thus implying that $(\mathrm{lsp}(S))^\perp \subseteq S^\perp$. We now prove the other inclusion: let $x \in S^\perp$ and take $z \in \mathrm{lsp}(S)$. Thus, $z = \sum_{i \in I} \lambda_i y_i$, with $y_i \in S$ and I finite set, and $(x, z) = (x, \sum_{i \in I} \lambda_i y_i) = \sum_{i \in I} \overline{\lambda_i}(x, y_i) = 0$. Hence $x \in \mathrm{lsp}(S)^\perp$.
  3. Take $x \in S \cap S^\perp$, then $(x, x) = 0 \implies x = 0$. This proves the third point of the lemma.

$\qquad\square$

Proposition 5.15 (Orthogonal decomposition) *Let M be a closed subspace of $\mathcal{H}$. Then for all $x \in \mathcal{H}$ there exist $x_1 \in M$ and $x_2 \in M^\perp$, uniquely determined, such that $x = x_1 + x_2$. Thus, $\mathcal{H} = M \oplus M^\perp$.*

Proof Take any $x \in \mathcal{H}$ and set $x_1 := P_x M$ and $x_2 = x - x_1$. We show that $x_2 \in M^\perp$, obtaining the decomposition claimed. We prove separately that $\mathrm{Re}(x_2, z) = 0$ and $\mathrm{Im}(x_2, z) = 0$, for any $z \in M$. Take any $z \in M$ and consider the vector $x_1 + tz$, with $t \in \mathbb{R}$. Then

$$\mathrm{d} := \mathrm{dist}(x, M) = \underbrace{\|x - x_1\|}_{=\|x_2\|} \leq \underbrace{\|x - (x_1 + tz)\|}_{=\|x_2 + tz\|},$$

impying that

$$\|x_2 + tz\|^2 = (x_2 + tz, x_2 + tz) \geq \|x_2\|^2, \quad \forall t \in \mathbb{R}.$$

Hence,

$$\|x_2\|^2 + t^2 \|z\|^2 + 2\,\mathrm{Re}(x_2, tz) \geq \|x_2\|^2 \implies t^2 \|z\|^2 + 2t\,\mathrm{Re}(x_2, z) \geq 0, \quad \forall t \in \mathbb{R},$$

implying that $\mathrm{Re}(x_2, z) = 0$.

Repeating the same argument replacing $(x_1 + tz)$ with $(x_1 + itz)$, we obtain $\mathrm{Im}(x_2, z) = 0$. This proves that $x_2 \in M^\perp$.

We are now left to prove uniqueness of the decomposition. Suppose that $x = x_1 + x_2 = y_1 + y_2$ with $x_1, y_1 \in M$ and $x_2, y_2 \in M^\perp$, then $M \ni x_1 - y_1 = y_2 - x_2 \in M^\perp$, implying that $x_1 = y_1$ and $x_2 = y_2$, thanks to point 3 of Lemma 5.14. $\qquad\square$

Lemma 5.16 *Let M be a subspace of $\mathcal{H}$. Then*

1. *$M^\perp$ is a closed subspace of $\mathcal{H}$;*
2. *$\overline{M}^\perp = M^\perp$;*
3. *$(M^\perp)^\perp = \overline{M}$;*
4. *$M^\perp = \{0\} \iff \overline{M} = \mathcal{H}$, i.e., M is dense in $\mathcal{H}$.*

Proof The first point is already proved by Lemma 5.14. Thus, we prove the other three points:

2. The first inclusion is immediate. Indeed, $M \subseteq \overline{M}$, implying that $\overline{M}^\perp \subseteq M^\perp$. To prove the other inclusion take any $x \in M^\perp$ and take any $z \in \overline{M}$: there exists a sequence $(z_n)_n$ in M such that $z = \lim_{n \to \infty} z_n$. Hence $(x, z) = \lim_{n \to \infty}(x, z_n) = 0$, so that $x \in \overline{M}^\perp$.
3. Write the decomposition of Proposition 5.15 with the sets $\overline{M}^\perp$ and $M^\perp$:

$$\mathcal{H} = \overline{M}^\perp \oplus \overline{M} = M^\perp \oplus \overline{M}, \quad \text{and} \quad \mathcal{H} = M^\perp \oplus (M^\perp)^\perp,$$

From uniqueness of the direct sum this proves that $(M^\perp)^\perp = \overline{M}$.

4. Using again the decomposition of Proposition 5.15 with $M^\perp$ and point 3 we have
$\mathcal{H} = M^\perp \oplus (M^\perp)^\perp = M^\perp \oplus \overline{M}$. Hence $M^\perp = \{0\}$ if and only if $\overline{M} = \mathcal{H}$.

$\square$

5.3 Basis of a Hilbert Space

Lemma 5.17 (Orthonomalization) *Let $\{u_i,\ i \in I \subseteq \mathbb{N}\}$ be a countable linearly independent subsystem of $\mathcal{H}$. Then there exists an orthonormal system $\{v_i, i \in I\}$ such that*

$$\mathrm{lsp}(\{v_i,\ i \in I\}) = \mathrm{lsp}(\{u_i,\ i \in I\}).$$

Proof We suppose that $I = \mathbb{N}$. The proof constructs the orthonormal system $\{v_i,\ i \in I\}$ explicitly, and in order to define the vector v_n, only the vectors $u_1, \ldots, u_n$ are needed. Thus, the result can be proved analogously for any subset $I \subset \mathbb{N}$.
Define iteratively

$$\begin{cases} w_1 = u_1, \\ v_1 = \frac{w_1}{\|w_1\|}, \end{cases} \quad \begin{cases} w_2 := u_2 - (u_2, v_1)v_1, \\ v_2 := \frac{w_2}{\|w_2\|}, \end{cases} \quad \cdots \quad \begin{cases} w_n := u_n - \sum_{k=1}^{n-1}(u_n, v_k)v_k, \\ v_n := \frac{w_n}{\|w_n\|}, \end{cases} \quad \forall n \in \mathbb{N}.$$

Clearly $\mathrm{lsp}(\{v_i,\ i = 1, \ldots, n\}) = \mathrm{lsp}(\{u_i,\ i = 1, \ldots, n\})$ for all $n \in \mathbb{N}$. Moreover, by construction, all the vectors v_i have norm equal to one. Thus, we are left to prove orthogonality of the vectors v_i to conclude the proof. We proceed by induction: Suppose that $(v_i, v_j) = \delta_{i,j}$ for all $1 \le i, j \le n$. Then we show that $(v_{n+1}, v_j) = 0$ for all $j = 1, \ldots, n$. Clearly it is equivalent to prove that $(w_{n+1}, v_j) = 0$ for all $j = 1, \ldots, n$. By construction we write, for any $j = 1, \ldots, n$:

$$(w_{n+1}, v_j) = \left(u_{n+1} - \sum_{k=1}^{n}(u_{n+1}, v_k)v_k, v_j\right)$$

$$= (u_{n+1}, v_j) - \sum_{k=1}^{n}(u_{n+1}, v_k)(v_k, v_j) = (u_{n+1}, v_j) - (u_{n+1}, v_j) = 0.$$

$$(5.7)$$

$\square$

Definition 5.18 (*Hilbert Basis*) Let $S \subseteq \mathcal{H}$ be a maximal orthonormal system, (i.e., there exists no orthonormal system $\tilde{S}$ such that $S \subset \tilde{S}$). S is said to be *complete*, or to be a Hilbert basis of $\mathcal{H}$.

Remark 5.19 An orthonormal system $S \subseteq \mathcal{H}$ is a basis if and only if $S^\perp = \{0\}$. Indeed if S were a basis and there exists $0 \ne z \in S^\perp$, then also $\tilde{S} = S \cup \{\frac{z}{\|z\|}\}$ would be a basis, contradicting the maximality of S. Viceversa if $S^\perp = \{0\}$, suppose by contradiction that there exists $\tilde{S}$ orthonormal system such that $S \subset \tilde{S}$, then there exists $z \in \tilde{S} \setminus S$, with $\|z\| = 1$, and $z \in S^{\perp}$, which contradicts Lemma 5.16.

Proposition 5.20 *Any Hilbert space has at least one basis.*

Proof Let $\mathcal{U} := \{S \subseteq \mathcal{H},\ S \text{ is an orthonormal system }\} \neq \emptyset$. Observe that the set $\mathcal{U}$ is ordered with respect to the inclusion $\subseteq$ and take $\mathcal{U}_1$ a totally ordered subsystem of $\mathcal{U}$ and define $\Sigma := \bigcup_{S \in \mathcal{U}_1} S$. Clearly Σ is an upper bound for $\mathcal{U}_1$, but using Zorn's Lemma we conclude that there exists a maximal element of $\mathcal{U}$ which turns out to be a basis for $\mathcal{H}$. $\qquad\square$

Proposition 5.21 *Let $\mathcal{H}$ be a Hilbert space. Then $\mathcal{H}$ admits a countable basis if and only if it is separable.*

Proof $\implies$) Let $S \subseteq \mathcal{H}$ be a countable basis and set $Y = \mathrm{lsp}(S)$. Then $Y^{\perp} = \mathrm{lsp}(S)^{\perp} = S^{\perp} = \{0\}$. Hence from Lemma 5.16 Y is dense in $\mathcal{H}$, consequently $\tilde{Y} := \mathrm{lsp}_{\mathbb{Q}}(S)$ is countable and dense. In particular we have $\mathcal{H} = \overline{\mathrm{lsp}(S)} = \overline{\mathrm{lsp}_{\mathbb{Q}}(S)}$.

$\impliedby$) Let $T := \{x_n,\ n \in \mathbb{N}\}$ be a countable and dense subset of $\mathcal{H}$ and construct a countable basis as follows: set $\tilde{x}_1 = x_{n_1}$, where x_{n_1} is the first non-zero element in T, next set $\tilde{x}_2 = x_{n_2}$ where $n_2 > n_1$ and x_{n_2} is the first element in $\{x_n,\ n > n_1\}$ such that $\{x_{n_1}, x_{n_2}\}$ is a linearly independent system. Iterate the procedure, defining $\tilde{x}_k = x_{n_k}$ where $n_k > n_{k-1}$ and x_{n_k} is the first element in $\{x_n,\ n > n_{k-1}\}$ such that $\{x_{n_1}, \ldots, x_{n_k}\}$ is a linearly independent system. Define $\tilde{T} := \{\tilde{x}_k, k \in \mathbb{N}\}$, clearly $\mathrm{lsp}(T) = \mathrm{lsp}(\tilde{T})$ and $\overline{T} = \mathcal{H}$ implies that $\overline{\mathrm{lsp}(\tilde{T})} = \mathcal{H}$. Apply the orthonormalization procedure of Lemma 5.17 to $\tilde{T}$, and obtain the Hilbert basis S. $\qquad\square$

Proposition 5.22 (Parseval identity) *Let $\mathcal{H}$ be a separable Hilbert space and let $S = \{e_i,\ i \in I \subseteq \mathbb{N}\}$ be a basis. Then for every $x \in \mathcal{H}$ we have:*

$$x = \sum_{i \in I}(x, e_i)e_i, \quad \|x\|^2 = \sum_{i \in I}|(x, e_i)|^2.$$

Proof Suppose without loss of generality that $I = \mathbb{N}$ and let $(y_n)_n$ be the sequence of partial sums $y_n := \sum_{i=1}^{n}(x, e_i)e_i$, so that $\|y_n\|^2 = \sum_{i=1}^{n}|(x, e_i)|^2$. We now prove that $y_n \to x$, showing first that $(y_n)_n$ converges to some limit point y and finally that $y = x$. Take $n, p \in \mathbb{N}$:

$$\|y_{n+p} - y_n\|^2 = \left\| \sum_{i=n+1}^{n+p}(x, e_i)e_i \right\|^2 = \sum_{i=1}^{n+p}|(x, e_i)|^2 \le \sum_{i=n+1}^{\infty}|(x, e_i)|^2,$$

since by Bessel inequality (5.4) $\sum_{i=1}^{\infty}|(x, e_i)|^2 \le \|x\|$, then

$$\sum_{i=n+1}^{\infty}|(x, e_i)|^2 \to 0, \quad \text{as } n \to \infty.$$

This proves that $(y_n)_n$ converges. Call $y = \lim_{n\to\infty} y_n$, we show that $y = x$: compute

$$(y_n - x, e_k) = \left(\sum_{i=1}^{n}(x, e_i)e_i, e_k\right) - (x, e_k) = (x, e_i)\delta_{ik} - (x, e_k) = 0, \ \forall n \geq k.$$

Since $y_n \to y$, passing to the limit in the above expression we obtain that $(y - x, e_k) = 0$ for all $k \in \mathbb{N}$, so that $y - x \in S^{\perp}$. Since S is a basis, $S^{\perp} = \{0\}$, giving $y = x$ and concluding the proof. $\square$

5.4 The Riesz Theorem

Theorem 5.23 (Riesz) *Let $\mathcal{H}$ be a Hilbert space. For every $T \in \mathcal{H}' = \mathcal{L}(\mathcal{H}, \mathbb{K})$ there exists a unique $y_T \in \mathcal{H}$ such that*

$$Tx = (x, y_T), \quad \forall x \in \mathcal{H}.$$

In addition, it holds $\|T\|_{\mathcal{H}'} = \|y_T\|_{\mathcal{H}}$.

Proof We first suppose existence of the vector y_T as claimed by the theorem and prove uniqueness, which is straightforward. Suppose that there exists $y_1, y_2 \in \mathcal{H}$ such that $Tx = (x, y_1) = (x, y_2)$ for all $x \in \mathcal{H}$. Then $(x, y_1 - y_2) = 0$ for all $x \in \mathcal{H}$, in particular choosing $x = y_1 - y_2$ we obtain that $\|y_1 - y_2\| = 0$ implying that $y_1 = y_2$.

We now turn to the proof of existence. If $T \equiv 0$ clearly the vector y_T exists and is $y_T = 0$. Suppose then that $T \neq 0$ and set $N := \ker(T) = \{x \in \mathcal{H} : Tx = 0\}$. From the continuity of T follows that N is a closed subspace of $\mathcal{H}$, thus, we may write $\mathcal{H} = N \oplus N^{\perp}$. Take $x_0 \in N^{\perp}$ (hence in particular $x_0 \neq 0$), we look for y_T of the form

$$y_T = \alpha x_0, \quad \text{for some } \alpha \in \mathbb{K}.$$

To this aim, we first pick an arbitrary $x \in \mathcal{H}$ and we write

$$x = \underbrace{\frac{Tx}{Tx_0}x_0}_{\in\, N^{\perp}} + \underbrace{\left(x - \frac{Tx}{Tx_0}x_0\right)}_{\in\, N}.$$

Next we impose $Tx = (x, \alpha x_0)$ and exploit this expression of x to find α:

$$Tx = (x, \alpha x_0) = \frac{Tx}{Tx_0}(x_0, \alpha x_0) \implies \alpha = \frac{\overline{Tx_0}}{\|x_0\|^2}.$$

Thus, defining $y_T := \frac{\overline{Tx_0}}{\|x_0\|^2}x_0$ the previous computations also show that $Tx = (x, y_T)$ for all $x \in \mathcal{H}$.

We conclude the proof showing the equality of the norms:

$$\|T\|_{\mathcal{H}'} = \sup_{\|x\|_{\mathcal{H}}=1} |Tx| = \sup_{\|x\|_{\mathcal{H}}=1} |(x, y_T)| = \|y_T\|_{\mathcal{H}}.$$

$\square$

Corollary 5.24 (Lax Milgram) *Let $\mathcal{H}$ be a Hilbert space and $\mathcal{B} : \mathcal{H} \times \mathcal{H} \to \mathbb{C}$ be a map satisfying the following properties:*

1. *$\mathcal{H} \ni x \to \mathcal{B}(x, y)$ is linear for any $y \in \mathcal{H}$;*
2. *$\mathcal{H} \ni y \to \mathcal{B}(x, y)$ is antilinear for any $x \in \mathcal{H}$;*
3. *there exists a non negative γ such that $|\mathcal{B}(x, y)| \leq \gamma \|x\| \|y\|$, for any $x, y \in \mathcal{H}$;*
4. *there exists $\delta > 0$ such that $\mathcal{B}(x, x) \geq \delta \|x\|^2$, for any $x \in \mathcal{H}$.*

Then there exists a unique $S \in \mathcal{L}(\mathcal{H})$ such that $(x, y) = \mathcal{B}(x, Sy)$, for all $x, y \in \mathcal{H}$. In particular, the following holds: for any $F \in \mathcal{H}'$ there exists a unique $y_F \in \mathcal{H}$ such that $Fx = \mathcal{B}(x, y_F)$, for all $x \in \mathcal{H}$.

Proof We start again by proving uniqueness of S. Suppose that there exist $S, \tilde{S} \in \mathcal{L}(\mathcal{H})$ such that

$$(x, y) = \mathcal{B}(x, Sy) = \mathcal{B}(x, \tilde{S}y), \quad \forall x, y \in \mathcal{H}.$$

It follows that $\mathcal{B}(x, (S - \tilde{S})y) = (x, 0) = 0$, for all $x \in \mathcal{H}$. Choosing $x = Sy - \tilde{S}y$ and using hypothesis 4 we obtain:

$$0 = \mathcal{B}(Sy - \tilde{S}y, Sy - \tilde{S}y) \geq \delta \|Sy - \tilde{S}y\|^2, \quad \forall y \in \mathcal{H}.$$

This implies that $S = \tilde{S}$.

Next we turn to the existence of S. To this aim, we define

$$\mathcal{D} := \{y \in \mathcal{H} : \exists\, y^* \in \mathcal{H} : (x, y) = B(x, y^*), \forall x \in \mathcal{H}\}$$

and we claim that $\mathcal{D} = \mathcal{H}$, so that defining $Sy := y^*$ we construct the required map S. First of all remark that $\mathcal{D}$ is non empty, since $0 \in \mathcal{D}$. Next we prove that $\mathcal{D} \ni y \to Sy$ is linear and continuous: let $y_1, y_2 \in \mathcal{D}$, then:

$$(x, y_1 + y_2) = (x, y_1) + (x, y_2) = \mathcal{B}(x, Sy_1) + \mathcal{B}(x, Sy_2) = \mathcal{B}(x, Sy_1 + Sy_2),$$

implying that $y_1 + y_2 \in \mathcal{D}$ and $S(y_1 + y_2) = Sy_1 + Sy_2$. Analogously, if $\lambda \in \mathbb{C}$ and $y \in \mathcal{D}$ then $\lambda y \in \mathcal{D}$, with $S(\lambda y) = \lambda Sy$. This proves linearity. To prove continuity we notice that, for every $y \in \mathcal{D}$

$$(Sy, y) = \mathcal{B}(Sy, Sy) \geq \delta \|Sy\|^2 \implies \delta \|Sy\|^2 \leq \|Sy\| \|y\| \implies \|Sy\| \leq \frac{1}{\delta} \|y\|. \quad (5.8)$$

Thus, we are left to prove that $\mathcal{D} = \mathcal{H}$. To this aim, we show that $\mathcal{D}$ is both closed and dense in $\mathcal{H}$ (obtaining that $\mathcal{D} = \overline{\mathcal{D}} = \mathcal{H}$).

- $\mathcal{D}$ **is closed:** let $(y_n)_n \in \mathcal{D}$, $y_n \to y$. By (5.8) we have $\|Sy_n - Sy_m\| \le \frac{1}{\delta}\|y_n - y_m\|$, thus, there exists a point $z \in \mathcal{H}$ such that $Sy_n \to z$. Moreover, for all $x \in \mathcal{H}$ we have $(x, y_n) = \mathcal{B}(x, Sy_n)$ and passing to the limit as n tends to infinity, we obtain $(x, y) = \mathcal{B}(x, z)$. Hence $\mathcal{D}$ is closed and $z = Sy$ with $y \in \mathcal{D}$.
- $\mathcal{D}$ **is dense:** we show that $D^{\perp} = \{0\}$. Suppose by contradiction that there exists a nonzero element $y_0 \in \mathcal{D}^{\perp}$. Then setting $Tx = \mathcal{B}(z, y_0)$, we see that $T \in \mathcal{H}'$ and, by the Riesz Theorem, that there exists a unique $y_T \in \mathcal{H}$ such that $Tz = \mathcal{B}(z, y_0) = (z, y_T)$, for all $z \in \mathcal{H}$. Thus, we obtain $y_T \in \mathcal{D}$ and $y_0 = Sy_T$. Since $y_0 \in \mathcal{D}^{\perp}$, we have

$$0 = (y_0, y_T) = \mathcal{B}(y_0, y_0) \ge \delta \, \|y_0\|^2 \, ,$$

obtaining a contradiction. Thus, $y_0 = 0$ and $\mathcal{D}$ is dense in $\mathcal{H}$.

Finally we show the last property of the statement: let $F \in \mathcal{H}'$, we know from the Riesz Theorem that there exists $\tilde{y}_F \in \mathcal{H}$ such that, for all $x \in \mathcal{H}$, we have $Fx = (x, \tilde{y}_F)$. But $(x, \tilde{y}_F) = B(x, S\tilde{y}_F)$, for all $x \in \mathcal{H}$, thus, $y_F = S\tilde{y}_F$, concluding the proof. $\qquad\qquad\square$

Proposition 5.25 *Any Hilbert space is reflexive.*

Proof Recall the definition of reflexivity: let $J : \mathcal{H} \to \mathcal{H}''$ be the map given by $\langle J(x), f \rangle_{\mathcal{H}'',\mathcal{H}'} = \langle f, x \rangle_{X',X}$. Then $\mathcal{H}$ is reflexive if J is surjective. We use the Riesz Theorem: define the surjective map

$$\mathrm{i} : \mathcal{H} \to \mathcal{H}', \quad \langle \mathrm{i}(y), x \rangle_{\mathcal{H}',\mathcal{H}} = (x, y), \qquad \text{for all } \ x, y \in \mathcal{H}.$$

Using previous notation, $y = y_T$ where $\mathrm{i}(y_T) = T$. The map i is an antilinear isometry, i.e.,

$$\mathrm{i}(\alpha y_1 + \beta y_2) = \overline{\alpha}\,\mathrm{i}(y_1) + \overline{\beta}\,\mathrm{i}(y_2), \quad \forall \alpha, \beta \in \mathbb{C}, \ \forall y_1, y_2 \in \mathcal{H},$$

and

$$\|\mathrm{i}(y)\|_{\mathcal{H}'} = \|y\|_{\mathcal{H}}, \quad \forall y \in \mathcal{H}.$$

Given $x, y \in \mathcal{H}$ we set $f = \mathrm{i}(x)$ and $g = \mathrm{i}(y)$. We define

$$[f, g] = [\mathrm{i}(x), \mathrm{i}(y)] = (y, x) = \overline{(x, y)}.$$

We prove that $[\cdot, \cdot]$ is a scalar product in $\mathcal{H}'$. It is easy to see that $[f, f] = \|x\|^2$, $[f_1 + f_2, g] = [f_1, g] + [f_2, g]$, $[\alpha f, g] = \alpha[f, g]$, and $\overline{[f, g]} = [g, f]$ verifying points $1 - 4$ of Definition 5.1. Moreover, by the Riesz Theorem, 5.23 $[f, f] = \|x\|_{\mathcal{H}}^2 = \|f\|_{\mathcal{H}'}^2$, proving that $(\mathcal{H}', [\cdot, \cdot])$ is a Hilbert space. Therefore we now apply the Riesz Theorem once again: there exists a surjective map $j : \mathcal{H}' \to \mathcal{H}''$ such that

$$\langle j(f), g \rangle_{\mathcal{H}'',\mathcal{H}'} = [g, f], \quad \forall f, g \in \mathcal{H}',$$

i.e.,

$$\langle j(f), g \rangle_{\mathcal{H}'',\mathcal{H}'} = [g, f] = [i(y), i(x)] = (x, y) = \langle i(y), x \rangle_{\mathcal{H}',\mathcal{H}} = \langle g, x \rangle_{\mathcal{H}',\mathcal{H}}.$$

Hence, $\langle (j \circ i)(x), g \rangle_{\mathcal{H}'',\mathcal{H}'} = \langle g, x \rangle_{\mathcal{H}',\mathcal{H}}$ and thus, setting $J = j \circ i$ we obtain a linear surjective isometry that satisfies the definition of reflexivity for $\mathcal{H}$. $\qquad\square$

5.5 The Adjoint Operator

Let $\mathcal{H}$ be a Hilbert space (over $\mathbb{K}$) and $T \in \mathcal{L}(\mathcal{H})$. Given $y \in \mathcal{H}$ arbitrary, consider the map

$$\mathcal{H} \ni x \xrightarrow{g} (Tx, y).$$

By linearity of T and of the scalar product with respect to the first variable, g is linear. In addition

$$|g(x)| = |(Tx, y)| \le \|Tx\| \, \|y\| \le \|T\| \, \|x\| \, \|y\|.$$

Hence $g \in \mathcal{H}'$ and using the Riesz Theorem there exists $y^* \in \mathcal{H}$ such that $g(x) = (x, y^*)$, for all $x \in \mathcal{H}$. We set $y^* =: T^*y$, with $T^* : \mathcal{H} \to \mathcal{H}$. From its definition it is an immediate check that T^* is linear. We compute its operatorial norm:

$$\|T^*y\| = \sup_{\|x\|=1} |(x, T^*y)| = \sup_{\|x\|=1} |(Tx, y)| \le \sup_{\|x\|=1} \|Tx\| \, \|y\| \le \sup_{\|x\|=1} \|T\| \, \|x\| \, \|y\|.$$

On the other hand:

$$\|Tx\| = \sup_{\|y\|=1} |(Tx, y)| = \sup_{\|x\|=1} |(x, T^*y)| \le \sup_{\|y\|=1} \|x\| \, \|T^*y\| = \|T^*\| \, \|x\|.$$

Thus, we have proved the following result.

Proposition 5.26 *Let $\mathcal{H}$ be a Hilbert space and $T \in \mathcal{L}(\mathcal{H})$. Then there exists a unique $T^* \in \mathcal{L}(\mathcal{H})$ such that*

$$(Tx, y) = (x, T^*y), \quad \forall x, y \in \mathcal{H}.$$

In addition $\|T^*\| = \|T\|$.

Definition 5.27 (*Adjoint operator and selfadjointness*) The operator T^* is called adjoint operator of T. If $T = T^*$, then T is called selfadjoint.

Exercise: let $\mathcal{H}$ be a Hilbert space, $F \subseteq \mathcal{H}$ a closed subspace and $P_F : \mathcal{H} \to \mathcal{H}$ the projection on F. Then $P \in \mathcal{L}(\mathcal{H})$, $\|P_F\| = 1$ and $P = P^*$.

Chapter 6
Spaces of Sequences

6.1 Spaces and Norms

Definition 6.1 (ℓ^p *spaces*) Let $1 \le p < \infty$. We define

$$\ell^p := \{a = (a_n)_n, \text{ sequence in } \mathbb{R} : \sum_{n=1}^{\infty} |a_n|^p < +\infty\}, \quad \|a\|_{\ell^p} := \|(a_n)_n\|_{\ell^p} = \Big(\sum_{n=1}^{\infty} |a_n|^p\Big)^{\frac{1}{p}}.$$

We will often write $\|a\|_p$ in place of $\|a\|_{\ell^p}$. For $p = \infty$ we define

$$\ell^{\infty} = \{a = (a_n)_n, \text{ sequence in } \mathbb{R} : \sup_{n \in \mathbb{N}} |a_n| < \infty\}, \quad \|a\|_{\ell^{\infty}} = \|(a_n)_n\|_{\ell^{\infty}} = \|a\|_{\infty} = \sup_{n \in \mathbb{N}} |a_n|.$$

Remark 6.2 Let $1 < p < q < \infty$. Then the following inclusions hold:

$$\ell^1 \subsetneq \ell^p \subsetneq \ell^q \subsetneq \ell^{\infty}.$$

Indeed let $a = (a_n)_n \in \ell^1$ and take $p > 1$. Since $\sum_{n=1}^{\infty} |a_n|$ is finite, $|a_n| \to 0$, and in particular there exists N_1 such that $|a_n| < 1$ for all $n > N_1$. Thus, $|a_n|^p < |a_n|$ for all $n > N_1$. Hence

$$\sum_{n=1}^{\infty} |a_n|^p = \sum_{n=1}^{N_1} |a_n|^p + \sum_{n=N_1+1}^{\infty} |a_n|^p \le \sum_{n=1}^{N_1} |a_n|^p + \|a\|_1 < \infty,$$

implying that $a \in \ell^p$. Analogously one proves that $\ell^p \subseteq \ell^q$. Let now $1 \le q < \infty$ and $a \in \ell^q$. Then $\sum_n |a_n|^q < \infty$ implies that there exists $N_2 > 0$ such that $|a_n| < 1$ for all $n > N_2$. Thus,

$$\sup_{n \in \mathbb{N}} |a_n| \le \sup_{n \le N_2} |a_n| + \sup_{n > N_2} |a_n| < \infty.$$

© The Author(s), under exclusive license to Springer Nature Switzerland AG 2026
S. Zagatti, *Functional Analysis*, SISSA Springer Series 7,
https://doi.org/10.1007/978-3-032-24475-8_6

Thus, $a \in \ell^\infty$.

Finally we remark that all inclusions are strict, indeed let $a = (a_n)_n$ with $a_n = \frac{1}{n}$, for all $n \in \mathbb{N}$. Then $a \notin \ell^1$ but $a \in \ell^p$ for all $p > 1$. Analogously taking $a = (a_n)_n$ with $a_n := \frac{1}{n^{\frac{1}{p}}}$ we obtain that $a \notin \ell^p$ but $a \in \ell^q$ for all $q > p$. Finally $a = (a_n)_n$ with $a_n = 1$ for all $n \in \mathbb{N}$ is not in ℓ^p for any finite p, but $a \in \ell^\infty$.

Now we define an operation and a norm on these sets to turn them into linear normed spaces: let $a = (a_n)_n$ and $b = (b_n)_n$ be two sequences in ℓ^p, $\lambda \in \mathbb{C}$ and define

$$a + b := (a_n + b_n)_n, \quad \lambda a := (\lambda a_n)_n.$$

It is immediate to check that if $p = 1$ or $p = \infty$, $a + b$ and λa are both sequences in ℓ^p. Take instead $p \neq 1$ and $p \neq \infty$. To prove that ℓ^p is a linear space with these operations we recall that, for two real numbers $\alpha, \beta \in \mathbb{R}$ we have:

$$|\alpha + \beta|^p \leq (2\max(|\alpha|, |\beta|))^p = 2^p \max(|\alpha|, |\beta|)^p = 2^p \max(|\alpha|^p, |\beta|^p) \leq 2^p(|\alpha|^p + |\beta|^p).$$

Thus, in our case

$$\sum_{n=1}^{\infty} |a_n + b_n|^p \leq 2^p \left(\sum_{n=1}^{\infty} |a_n|^p + \sum_{n=1}^{\infty} |b_n|^p \right) < \infty.$$

Obviously $\sum_n |\lambda a_n|^p \leq |\lambda|^p \sum_n |a_n|^p$, thus proving that ℓ^p is a linear space for any $1 \leq p \leq \infty$. Is now $(\ell^p, \|\cdot\|_p)$ a normed space? If $p = 1$ or $p = \infty$ then it is immediate to check that $\|\cdot\|_{\ell^1}$ and $\|\cdot\|_{\ell^\infty}$ are norms, thus, we leave this as an exercise. Instead we prove this result in the more interesting case $1 < p < \infty$. To this aim, we need the following preliminary lemma:

Lemma 6.3 (Young inequality) *Let $p, q \in]1, \infty[$ be such that $\frac{1}{p} + \frac{1}{q} = 1$. Then*

$$\alpha\beta \leq \frac{\alpha^p}{p} + \frac{\beta^q}{q}, \quad \forall \alpha, \beta \geq 0.$$

Proof Suppose that $\alpha, \beta > 0$ and consider the map $]0, \infty[\ni t \to \log t$, which is a *concave map*, i.e.,

$$\log((1-t)x + ty) \geq (1-t)\log x + t\log y, \quad \forall x, y > 0, \text{ and } \forall t \in [0, 1].$$

Hence

$$\log\left(\frac{1}{p}\alpha^p + \frac{1}{q}\beta^q\right) \geq \frac{1}{p}\log\alpha^p + \frac{1}{q}\log\beta^q = \log\alpha + \log\beta = \log(\alpha\beta).$$

Taking the exponential on both sides we obtain

$$\frac{1}{p}\alpha^p + \frac{1}{q}\beta^q \geq \alpha\beta.$$

$\square$

Proposition 6.4 *Let* $1 < p < \infty$. *Then* $(\ell^p, \|\cdot\|_{\ell^p})$ *is a normed space.*

Proof Let a in ℓ^p. Then

1. $\|a\|_{\ell^p}^p = \sum_n |a_n|^p = 0 \iff a = 0$, i.e., $a_n = 0$ for all $n \in \mathbb{N}$;
2. $\|\lambda a\|_{\ell^p} = |\lambda| \, \|a\|_{\ell^p}$.

We are left to prove the triangular inequality: take $q > 1$ such that $\frac{1}{p} + \frac{1}{q} = 1$, i.e., $q = \frac{p}{p-1}$. Then, for any $a \in \ell^p$, $a \neq 0$ and $b \in \ell^q$, $b \neq 0$, we compute, using the Young inequality:

$$\frac{|a_n|}{\|a\|_p} \frac{|b_n|}{\|b\|_q} \leq \frac{1}{p} \frac{|a_n|^p}{\|a\|_p^p} + \frac{1}{q} \frac{|b_n|^q}{\|b\|_q^q}, \quad \forall n \in \mathbb{N}.$$

Thus, summing over $n \in \mathbb{N}$ we obtain:

$$\frac{1}{\|a\|_p \|b\|_q} \sum_{n=1}^{\infty} |a_n||b_n| \leq \frac{1}{p} \frac{\sum |a_n|^p}{\|a\|_p^p} + \frac{1}{q} \frac{\sum |b_n|^q}{\|b\|_q^q} = \frac{1}{p} + \frac{1}{q} = 1.$$

Hence we have the following inequality, known as the *Hölder inequality*:

$$\sum_{n=1}^{\infty} |a_n||b_n| \leq \|a\|_p \|b\|_q .$$

We now conclude the proof of the triangular inequality using the Hölder inequality:

$$\sum_{n=1}^{\infty} |a_n + b_n|^p = \sum_{n=1}^{\infty} |a_n + b_n|^{p-1} |a_n + b_n|$$

$$\leq \sum_{n=1}^{\infty} |a_n + b_n|^{p-1} |a_n| + \sum_{n=1}^{\infty} |a_n + b_n|^{p-1} |b_n|. \qquad (6.1)$$

Using now the Hölder inequality we obtain:

$$S = \sum_{n=1}^{\infty} |a_n + b_n|^{p-1} |a_n| \leq \left(\sum_{n=1}^{\infty} (|a_n + b_n|^{p-1})^q \right)^{\frac{1}{q}} \left(\sum_{n=1}^{+\infty} |a_n|^p \right)^{\frac{1}{p}},$$

but

$$(|a_n + b_n|^{p-1})^q = (|a_n + b_n|^{p-1})^{\frac{p}{p-1}} = |a_n + b_n|^p,$$

hence $S \leq (\sum |a_n + b_n|^p)^{\frac{p-1}{p}} (\sum |a_n|^p)^{\frac{1}{p}}$. Repeating the same computations for the other term in (6.1) we obtain:

$$\sum_{n=1}^{+\infty} |a_n + b_n|^p \leq \left(\sum_{n=1}^{\infty} |a_n + b_n|^p\right)^{1-\frac{1}{p}} [(\sum_{n=1}^{+\infty} |a_n|^p)^{\frac{1}{p}} + (\sum_{n=1}^{+\infty} |b_n|^p)^{\frac{1}{p}}].$$

Hence $\|a + b\|_p = (\sum_{n=1}^{+\infty} |a_n + b_n|^p)^{\frac{1}{p}} \leq \|a\|_p + \|b\|_p$ which proves the triangular inequality. $\square$

6.2 Main Properties

Completeness

We now pass to completeness, proving that $(\ell^p, \|\cdot\|_p)$ is a Banach space for all $p \in [1, \infty]$. We prove the case $p = 1$ for simplicity in the calculation, but the proof proceeds in the exact same way for the case $1 < p < \infty$, thus, we omit it. Then we prove separately the case $p = \infty$.

Proposition 6.5 *The normed space $(\ell^1, \|\cdot\|_1)$ is a Banach space.*

Proof Let $(a^n)_n$ be a Cauchy sequence in ℓ^1, where we denote by $a^n = (a_k^n)_{k \in \mathbb{N}}$, an element of the sequence, which is in turn a sequence in ℓ^1. Since $\|a^n - a^m\|_1 = \sum_{k=1}^{\infty} |a_k^n - a_k^m|$ tends to zero as $n, m \to \infty$, we clearly have that for every $h \in \mathbb{N}$, $|a_h^n - a_h^m| \leq \sum_{k=1}^{\infty} |a_k^n - a_k^m| \to 0$ as $n, m \to \infty$. Thus, $(a_h^n)_n$ is a Cauchy sequence in $\mathbb{R}$, and then there exists $a_h \in \mathbb{R}$ such that $a_h^n \to a_h$. Set $a = (a_h)_h$ and observe that

$$\sum_{k=1}^{\infty} |a_k^n - a_k| = \lim_{m \to \infty} \sum_{k=1}^{\infty} |a_k^n - a_k^m|.$$

Moreover, given $\varepsilon > 0$ there exists $N_\varepsilon \in \mathbb{N}$ such that $\sum_k |a_k^n - a_k^m| \leq \varepsilon$, for all $n, m \geq N_\varepsilon$, and thus,

$$\sum_k |a_k^n - a_k| \leq \varepsilon, \quad \forall n \geq N_\varepsilon, \quad \text{i.e. } a^n \to a, \text{ in } \ell^1.$$

$\square$

Proposition 6.6 *Let $1 < p < \infty$. The normed space $(\ell^p, \|\cdot\|_p)$ is a Banach space.*

Proof Omitted: adapt the case $p = 1$. $\square$

Proposition 6.7 *The normed space $(\ell^\infty, \|\cdot\|_\infty)$ is a Banach space.*

Proof Let $(a^n)_n$ be a Cauchy sequence in ℓ^∞, i.e., $\sup_{k\in\mathbb{N}} |a_k^n - a_k^m| \to 0$ as $n, m \to \infty$. Reasoning as in the previous proof we see that, for every fixed $h \in \mathbb{N}$, the real sequence $(a_h^n)_{n\in\mathbb{N}}$ is a Cauchy sequence in $\mathbb{R}$, thus, it converges to a limit value a_h. Setting $a = (a_k)_{k\in\mathbb{N}}$, we conclude since $|a_k^n - a_k| = \lim_{m\to\infty} |a_k^n - a_k^m|$ and for all positive ε there exists $N_\varepsilon \in \mathbb{N}$ such that $\sup_{k\in\mathbb{N}} |a_k^n - a_k^m| \le \varepsilon$ for all $n, m \ge N_\varepsilon$:

$$\sup_k |a_k^n - a_k| \le \varepsilon \quad \forall n \ge N_\varepsilon, \text{ implying that } a^n \to a \text{ in } \ell^\infty.$$

$\square$

Separability

Now we turn to the last property that we prove in these lecture notes regarding ℓ^p spaces, which is separability. We first prove that ℓ^p is separable if $p \ne \infty$, constructing explicitly a countable and dense subset of ℓ^p. Finally we show that ℓ^∞ is not separable. We start defining the following sets: for any $N \in \mathbb{N}$ let

$$S_N := \{a = (a_n)_n, a_n \in \mathbb{R}, a_n = 0 \ \forall n \ge N + 1\}.$$

Clearly $S_N \subseteq \ell^p$ for all $p \in [1, \infty)$ and for all $N \in \mathbb{N}$. Moreover, S_N is isomorphic to $\mathbb{R}^N$. Let now $a = (a_n)_n \in \ell^1$ and fix an arbitrary $\varepsilon > 0$, there exists $N_\varepsilon \in \mathbb{N}$ such that $\sum_{n=N_\varepsilon+1}^\infty |a_n| \le \varepsilon$. Thus, defining

$$b_n = \begin{cases} a_n & \text{if } n \le N_\varepsilon, \\ 0 & \text{if } n \ge N_\varepsilon + 1 \end{cases}$$

and considering the sequence $b = (b_n)_n \in S_{N_\varepsilon} \subseteq \ell^1$ we have $\|a - b\|_{\ell^1} = \sum_{n=N_\varepsilon+1}^\infty |a_n| \le \varepsilon$. Hence $\bigcup_{N\in\mathbb{N}} S_N$ is dense in ℓ^1, and considering $S_N(\mathbb{Q}) := \{a = (a_n)_n \in S_N, a_n \in \mathbb{Q} \ \forall n\}$, so that $S_N(\mathbb{Q})$ is isomorphic to $\tilde{\mathbb{Q}}^N$, we have that $\bigcup_{N\in\mathbb{N}} S_N(\mathbb{Q})$ is dense and countable in ℓ^1, thus, ℓ^1 is separable.

The same kind of argument proves separability of ℓ^p for any $p \ne \infty$.

Now we show that ℓ^∞ is not separable. To this aim, we need the following general lemma.

Lemma 6.8 *Let (X, d) be a separable metric space and let $\mathcal{E}$ be a family of non-empty open and pairwise disjoint subsets of X. Then $\mathcal{E}$ is countable.*

Proof Let $S \subseteq X$ be a countable and dense subset in X and let $\varphi : \mathbb{N} \to S$ be a bijection. Given $J \in \mathcal{E}$, let $n = n(J)$ be the first natural number such that $\varphi(n) \in J$. This defines a map $\mathcal{E} \ni J \to n(J) \in \mathbb{N}$, which is injective (since if $n(I) = m(J)$, then $\varphi(n) = \varphi(m) \in I \cap J$, implying $I = J$ because either I and J are disjoint or $I = J$) and thus, implies that $\mathcal{E}$ is countable. $\square$

Corollary 6.9 *The space ℓ^∞ is not separable.*

Proof We prove the corollary using previous lemma and reasoning by contradiction. Consider the family $\mathcal{E} = \{a \in (a_n)_n \in \ell^\infty, \ a_n \in \{0, 1\}\}$ and let $\tilde{\mathcal{E}} = \{E_a = B_{\ell^\infty}(a, \frac{1}{2}) : \ a \in \mathcal{E}\}$. Clearly, if $a \neq b$, then $E_a \cap E_b = \emptyset$, since $\|a - b\|_\infty = 1$. Hence $\tilde{\mathcal{E}}$ satisfies the requirements of previous proposition. Assume then by contradiction that ℓ^∞ is separable, then $\tilde{\mathcal{E}}$ is countable from previous lemma, but $\mathcal{E}$ is in bijection with $\mathbb{R}$ since $2^\mathbb{N} = \mathbb{R}$. $\qquad\square$

Chapter 7
L^p Spaces

In this section, we study L^p spaces: we start by defining them, but we omit the proofs of the first general results which are supposed to be known and for which we refer to [1].

7.1 Preliminaries and Definition of L^p Spaces

Let $\Omega \subseteq \mathbb{R}^d$ be an open set and consider the real linear space of continuous functions over Ω, $\mathcal{C}(\Omega) := \mathcal{C}(\Omega, \mathbb{R})$, and the subset

$$\mathcal{C}_b(\Omega) := \{u \in \mathcal{C}(\Omega) : \sup_\Omega |u| < \infty\}. \tag{7.1}$$

We remark that $\mathcal{C}_b(\Omega)$ is a subspace of $\mathcal{C}(\Omega)$, and we set $\|u\|_{\mathcal{C}_b(\Omega)} = \|u\|_\infty = \sup_\Omega |u|$. $(\mathcal{C}_b(\Omega), \|\cdot\|_\infty)$ is a Banach space.

Definition 7.1 Given $u \in \mathcal{C}(\Omega)$, consider the family $\mathcal{U} := \{O \subseteq \Omega : O \text{ open}, u|_O \equiv 0\}$. Then the set

$$\mathcal{N} := \bigcup_{O \in \mathcal{U}} O$$

is the *domain of nullity* of u and

$$\operatorname{supp}(u) := \Omega \setminus \mathcal{N} \tag{7.2}$$

is the *support* of u.

Notice that $\operatorname{supp}(u)$ is a closed set in Ω, and that

$$\operatorname{supp}(u) = \overline{\{x \in \Omega : u(x) \neq 0\}}^{\Omega}.$$

© The Author(s), under exclusive license to Springer Nature Switzerland AG 2026
S. Zagatti, *Functional Analysis*, SISSA Springer Series 7,
https://doi.org/10.1007/978-3-032-24475-8_7

Definition 7.2 We define the set of *compactly supported* functions as

$$\mathcal{C}_c(\Omega) = \{u \in \mathcal{C}(\Omega) \; : \; \mathrm{supp}(u) \text{ is a compact subset of } \Omega\}. \tag{7.3}$$

Remark 7.3 We collect here the following properties:

1. If $u \in \mathcal{C}_c(\Omega)$, then there exists $\delta_u > 0$ such that $\mathrm{dist}(\mathrm{supp}(u), \Omega^c) = \delta > 0$;
2. $\mathcal{C}_c(\Omega)$ is a subspace of $\mathcal{C}_b(\Omega)$;
3. $\mathcal{C}_c(\Omega)$ is *not* closed in $\mathcal{C}_b(\Omega)$.

Definition 7.4 We set $\mathcal{C}_0(\Omega) = \overline{\mathcal{C}_c(\Omega)}$ in $\mathcal{C}_b(\Omega)$, where the closure is taken in the infinity norm.

Remark 7.5 Let $u \in \mathcal{C}_0(\Omega)$. If $x_0 \in \partial\Omega$, then $\lim_{x \to x_0} u(x) = 0$. Similarly if Ω is an unbounded set and $(x_n)_n$ is a sequence in Ω, such that $|x_n| \to \infty$, then $\lim_{n \to \infty} u(x_n) = 0$.

From now on we will use some notions and definitions of measure theory. The measure that we will refer to is the Lebesgue measure.

Definition 7.6 (L^p *spaces*) Let $E \subseteq \mathbb{R}^d$ be a measurable set and let $1 \le p \le \infty$. If $p \ne \infty$ we set

$$\mathcal{L}^p(E) := \{f : E \to \mathbb{R}, \; f \text{ measurable such that } \int_E |f|^p \mathrm{d}x < \infty\}.$$

If $p = \infty$ we set

$$\mathcal{L}^\infty(E) := \{f : E \to \mathbb{R}, \; f \text{ measurable such that } \mathrm{ess}\sup_E |f| < \infty\}.$$

Given $f, g \in \mathcal{L}^p(E)$ we say that $f \sim g$ if $f(x) = g(x)$ for almost every x in E. We remark that $\sim$ is an equivalence relation in $\mathcal{L}^p(E)$ and we set, for $1 \le p \le \infty$,

$$L^p(E) = \mathcal{L}^p(E)/_\sim.$$

A function f in the space $L^p(E)$ is identified with its equivalence class. Roughly speaking we can think of f as a family of functions which coincide almost everywhere, or equivalently as a function which can be modified over any Lebesgue negligible set.

Definition 7.7 (L^p *norms*) Given $f \in L^p(E)$ we set

$$\|f\|_p = \|f\|_{L^p(E)} = \left(\int_E |f|^p \mathrm{d}x \right)^{\frac{1}{p}}, \quad \text{if } 1 \le p < \infty,$$

and

$$\|f\|_\infty = \|f\|_{L^\infty(E)} = \mathrm{ess}\sup_E |f|.$$

Finally if $p = 2$, f, g in $L^2(E)$ we set

$$(f, g)_{L^2(E)} = \int_E fg\,dx. \tag{7.4}$$

We remark that in case of complex valued functions (which are measurable if both the real and the imaginary parts are as real valued functions) all definitions remain the same, except for (7.4), that is set as

$$(f, g)_{L^2(E)} = \int_E f\overline{g}\,dx.$$

We state the following result without proving it and referring to [1, Theorem 4.8].

Proposition 7.8 $(L^p(E), \|\cdot\|_p)$ *is an Banach space for all* $1 \le p \le \infty$. $(L^2(E), \|\cdot\|_2)$ *is a Hilbert space.*

We just remark that, in order to prove Proposition 7.8, one needs to use Hölder inequality, which is a fundamental inequality and thus, we state it separately here.

<u>Notation:</u> when a pair of indices $p, q \in [1, \infty]$ is such that $\frac{1}{p} + \frac{1}{q} = 1$ (where we set $q = \infty$ if $p = 1$ and $q = 1$ if $p = \infty$), we say that (p, q) are *conjugated indices* and we write $q = p'$ and $p = q'$.

Lemma 7.9 (Hölder inequality) *Let* $p, q \in [1, \infty]$ *be conjugate indices. Then for every* f *in* $L^p(E)$ *and* g *in* $L^q(E)$ *we have*

$$\int_E |fg|\,dx \le \|f\|_p \|g\|_q.$$

Remark that if $p = q = 2$, the Hölder inequality is the Schwarz inequality for the Hilbert space $L^2(E)$.

Proposition 7.10 *Let* $\Omega \subseteq \mathbb{R}^d$ *be an open set and let* $1 \le p < \infty$. *The set* $C_c(\Omega)$ *is dense in* $L^p(\Omega)$.

We remark that the same result is false if $p = \infty$. We omit the proof of this proposition and we just prove the following important consequence.

Corollary 7.11 *Let* $\Omega \subseteq \mathbb{R}^d$ *be an open set and let* $1 \le p < \infty$. *Then* $L^p(\Omega)$ *is separable.*

Proof For every natural number n define the set

$$\Omega_n := \left\{ x \in \Omega : \operatorname{dist}(x, \Omega^c) > \frac{1}{n} \right\} \bigcap B(0, n).$$

Clearly we have $\Omega = \bigcup_{n \in \mathbb{N}} \Omega_n$ and, for every n, Ω_n is a bounded open set. Let now $f \in L^p(\Omega)$ and fix $\varepsilon > 0$ arbitrary. Then using previous proposition there exists

$\varphi \in C_c(\Omega)$ such that $\|f - \varphi\|_p \leq \varepsilon$. Moreover, there also exists $\bar{n} \in \mathbb{N}$ such that $\operatorname{supp}(\varphi) \subseteq \Omega_{\bar{n}}$. Hence $\varphi|_{\Omega_{\bar{n}}} \in C(\Omega_{\bar{n}})$, and $\Omega_{\bar{n}}$ is relatively compact in $\mathbb{R}^d$. Let now $q(\cdot)$ be a real polynomial with rational coefficients such that $\|\varphi - q\|_{C(\Omega_{\bar{n}})} \leq \dfrac{\varepsilon}{m(\Omega_{\bar{n}})^{\frac{1}{p}}}$, (where $m(A)$ is the Lebesgue measure of the set A). This choice of $q(\cdot)$ is possible using the Stone-Weierstrass Theorem 4.12. We set

$$\tilde{q} = \begin{cases} q & \text{on } \Omega_{\bar{n}} \\ 0 & \text{on } \Omega \setminus \Omega_{\bar{n}}, \end{cases}$$

i.e., $\tilde{q} = q\chi_{\Omega_{\bar{n}}}$, with $\chi_{\Omega_{\bar{n}}}$ the characteristic function of the set $\Omega_{\bar{n}}$. We compute

$$\|f - \tilde{q}\|_p \leq \|f - \varphi\|_p + \|\varphi - \tilde{q}\|_p \leq 2\varepsilon,$$

since

$$\|\varphi - \tilde{q}\|_p^p = \int_\Omega |\varphi - \tilde{q}|^p \mathrm{d}x = \int_{\Omega_{\bar{n}}} |\varphi - \tilde{q}|^p \mathrm{d}x \leq \int_{\Omega_{\bar{n}}} \frac{\varepsilon^p}{(m(\Omega_{\bar{n}})^{\frac{1}{p}})^p} \mathrm{d}x = \varepsilon^p. \quad (7.5)$$

Thus, considering the family $\mathcal{Q} := \{q\chi_{\Omega_n} : n \in \mathbb{N}, q \text{ polynomial with rational coefficients}\}$, it is countable and dense in $L^p(\Omega)$. $\qquad\qquad\square$

We conclude this introductory section showing that, if the measurable set E has finite measure, then the following inclusion holds:

Proposition 7.12 *Let $E \subseteq \mathbb{R}^d$ be a measurable set of finite measure $m(E)$ and let $1 \leq p \leq q \leq \infty$. Then $L^q(E) \subseteq L^p(E)$.*

Proof Let $q = \infty$ and $p < \infty$. Then for any $f \in L^\infty$ we have

$$\int_E |f|^p \mathrm{d}x \leq \int_E \|f\|_\infty^p \, \mathrm{d}x = \|f\|_\infty^p \, m(E).$$

Hence $f \in L^p(E)$.

Let now $1 \leq p \leq q < \infty$. For any $f \in L^q(E)$ we have that

$$\begin{aligned}
\int_E |f|^p \mathrm{d}x &= \int_E 1 |f|^p \mathrm{d}x \\
&\overset{7.9}{\leq} \left(\int_E (|f|^p)^{\frac{q}{p}} \mathrm{d}x \right)^{\frac{p}{q}} \left(\int_E 1^{(\frac{q}{p})'} \mathrm{d}x \right)^{((\frac{q}{p})')^{-1}} \\
&= \left(\int_E |f|^q \mathrm{d}x \right)^{\frac{p}{q}} \left(\int_E \mathrm{d}x \right)^{((\frac{q}{p})')^{-1}}.
\end{aligned} \quad (7.6)$$

And computing $\left(\frac{q}{p}\right)'$ we obtain

$$\left(\frac{q}{p}\right)' = \frac{q}{p}\frac{p}{q-p} = \frac{q}{q-p},$$

hence

$$\int_E |f|^p \mathrm{d}x \le \left(\int_E |f|^q \mathrm{d}x\right)^{\frac{p}{q}} m(E)^{\frac{q-p}{q}},$$

proving that

$$\|f\|_p \le \|f\|_q \, m(E)^{\frac{q-p}{pq}}.$$

$\square$

7.2 Convolution Product

In this section, we define the convolution product between two functions, which is a particularly significant operation in L^p spaces, since it satisfies nice properties that we will state and prove. We start with some preliminary notations and definitions needed.

Let $f : \mathbb{R}^d \to \overline{\mathbb{R}}$. We introduce the following notation:

1. $\check{f}(x) = f(-x), \quad x \in \mathbb{R}^d$;
2. given any $a \in \mathbb{R}^d$, $(\tau_a f)(x) := f(x - a), \quad x \in \mathbb{R}^d$.

Remark that, for any $a, b \in \mathbb{R}^d$ we have $\tau_a \check{f} = (\tau_{-a} f)\check{}$, and $\tau_a(\tau_b f) = \tau_{a+b} f$. Moreover, it is immediate from this definition that for any $p \in [1, \infty]$ and $f \in L^p(\mathbb{R}^d)$, $\left\|\check{f}\right\|_p = \|f\|_p$ and given any $a \in \mathbb{R}^d$, $\|\tau_a f\|_p = \|f\|_p$.

Lemma 7.13 *Let* $1 \le p < \infty$ *and* $f \in L^p$. *The map* $\mathbb{R}^d \ni a \mapsto \tau_a f \in L^p$ *is uniformly continuous.*

Proof We shall prove that for any $\varepsilon > 0$ there exists a corresponding positive δ such that for any $a, b \in \mathbb{R}^d$ with euclidean distance smaller than δ, then $\|\tau_a f - \tau_b f\|_p \le \varepsilon$. To this aim, we first show that we can, without loss of generality, suppose that $b = 0$. Indeed, since $(\tau_a f)(x) - (\tau_b f)(x) = f(x - a) - f(x - b)$ we have

$$\|\tau_a f - \tau_b f\|_p^p = \int |f(x - a) - f(x - b)|^p \mathrm{d}x = \int |f(y - (a - b)) - f(y)|^p \mathrm{d}x$$

$$= \|\tau_{(a-b)} f - f\|_p^p.$$

$$(7.7)$$

Thus, we can consider $b = 0$ and reduce ourselves to prove that there exists $\delta > 0$ such that $\|\tau_a f - f\|_p \le \varepsilon$ for all $|a| < \delta$. To do so we use density of $C_c(\mathbb{R}^d)$ in $L^p(\mathbb{R}^d)$. Indeed let $f \in C_c(\mathbb{R}^d)$, since f is uniformly continuous, there exists $\delta > 0$

such that $|f(x) - f(y)| \leq \varepsilon$, for all $|x - y| \leq \delta$. Thus, in particular, if $|a| < \delta$, we have

$$\int |f(x - a) - f(x)|^p \mathrm{d}x \leq \varepsilon^p m(A),$$

where $A := \{x \in \mathbb{R}^d \ : \ f(x - a) - f(x) \neq 0\}$ and clearly $m(A) \leq 2m(\mathrm{supp}(f))$ because $A \subset \mathrm{supp}(f) \cup (\mathrm{supp}(f) + a)$, hence $\|\tau_a f - f\|_p \leq 2^{\frac{1}{p}} m(\mathrm{supp}(f))^{\frac{1}{p}} \varepsilon$.

Finally for a general $f \in L^p$, take $(f_n)_n$ in $C_c(\mathbb{R}^d)$ such that f_n converges to f in L^p and write

$$\|\tau_a f - f\|_p = \|\tau_a f - \tau_a f_n\|_p + \|\tau_a f_n - f_n\|_p + \|f_n - f\|_p . \tag{7.8}$$

Then clearly $\tau_a f_n$ converges to $\tau_a f$ in L^p and we can find $\bar{n}$ sufficiently large such that $\|f_{\bar{n}} - f\|_p < \varepsilon$ and $\|\tau_a f_{\bar{n}} - \tau_a f\|_p < \varepsilon$. Thus, choosing $\delta > 0$ such that $\|\tau_a f_{\bar{n}} - f_{\bar{n}}\|_p < \varepsilon$ and using (7.8) we obtain $\|\tau_a f - f\|_p < 3\varepsilon$, for all $|a| < \delta$. $\square$

We can finally define the convolution product between two measurable functions:

Definition 7.14 Let $f, g : \mathbb{R}^d \to \mathbb{R}$ be two measurable functions. We say that f and g are *convolvable* if

$$(\check{\tau_x} f)(y) g(y) \in L^1(\mathbb{R}^d), \quad \text{for almost every } x \in \mathbb{R}^d.$$

In such case we set, for almost every $x \in \mathbb{R}^d$:

$$\begin{aligned}
(f * g)(x) &= \int (\check{\tau_x} f)(y) g(y) \mathrm{d}y = \int f(x - y) g(y) \mathrm{d}y \\
&= \int f(y) g(x - y) \mathrm{d}y = \int (\tau_x g)(y) f(y) \mathrm{d}y = (g * f)(x).
\end{aligned} \tag{7.9}$$

We call $f * g = g * f$ the *convolution product* of f and g.

We start proving the first properties of the convolution product:

Lemma 7.15 *Let f and g be convolvable. Then*

$$\mathrm{supp}(f * g) \subseteq \overline{\mathrm{supp}(f) + \mathrm{supp}(g)}.$$

Proof Let $x \in \overline{\mathrm{supp}(f) + \mathrm{supp}(g)}^c$, then $x \in \overline{(\mathrm{supp}(f) + \mathrm{supp}(g))^c}$. Thus, in particular $(x - \mathrm{supp}(f)) \cap \mathrm{supp}(g) = \emptyset$. But since $x - \mathrm{supp}(f) = \mathrm{supp}(\check{\tau_x} f)$ this implies that $(f * g)(x) = 0$, proving the claim. $\square$

Proposition 7.16 *Let $p, q \in [1, \infty]$ be a pair of conjugated indices and take $f \in L^p$ and $g \in L^q$. Then*

1. *f and g are convolvable;*
2. *$(f * g) \in C_b(\mathbb{R}^d)$ and $\|f * g\|_\infty \leq \|f\|_p \|g\|_p$;*
3. *if $1 < p, q < \infty$, then $f * g \in C_0(\mathbb{R}^d)$.*

Proof 1. $|(f * g)(x)| \leq \int |(\tau_x \check{f})g| dx \leq \|\tau_x \check{f}\|_p \|g\|_q = \|f\|_p \|g\|_q$, for all $x \in \mathbb{R}^d$.

2. The boundedness of $(f * g)$ follows from point 1, thus, we just check continuity of $f * g$, taking $x, x' \in \mathbb{R}^d$ and computing, for $p \neq \infty$, $(f * g)(x) - (f * g)(x')$:

$$|(f * g)(x) - (f * g)(x')| = |\int (\tau_x \check{f} - \tau_{x'} \check{f}) g \, dy| \leq \int |\tau_x \check{f} - \tau_{x'} \check{f}| |g| \, dy$$

$$\leq \left\| \tau_x \check{f} - \tau_{x'} \check{f} \right\|_p \|g\|_q.$$

$$(7.10)$$

This proves continuity for $p \neq \infty$ because of Lemma 7.13. Since $\mathbb{R}^d \ni x \to \tau_x f \in L^p$ is uniformly continuous just for $p \neq \infty$, to tackle the case $p = \infty$ it is enough to notice that $(f * g)(x) - (f * g)(x') = \int (\tau_x \check{g} - \tau_{x'} \check{g}) f \, dy$ and repeat the same computation above.

3. We use the definition of $C_0(\mathbb{R}^d)$ and we look for a sequence $(\varphi_n)_n$ in $C_c(\mathbb{R}^d)$ such that $\varphi_n \to f * g$ uniformly over $\mathbb{R}^d$. Since $1 < p, q < \infty$, there exist two sequences $(f_n)_n$ and $(g_n)_n$ in $C_c(\mathbb{R}^d)$ such that $(f_n)_n$ converges to f in L^p and $(g_n)_n$ converges to g in L^q. Thus, we consider the sequence $(\varphi_n)_n = (f_n * g_n)_n$, and use point 2: $(f_n * g_n)_n$ is a sequence in $C_b(\mathbb{R}^d)$ and using Lemma 7.15 we have that, since the sum of compact supports is precompact, then $f_n * g_n \in C_c(\mathbb{R}^d)$, for all $n \in \mathbb{N}$. We conclude the proof computing:

$$|(f_n * g_n)(x) - (f * g)(x)| \leq |(f_n * g_n)(x) - (f_n * g)(x)| + |(f_n * g)(x) - (f * g)(x)|$$

$$= \underbrace{|(f_n * (g_n - g))(x)|}_{\leq \|f_n\|_p \|g_n - g\|_q \to 0} + \underbrace{|((f_n - f) * g)(x)|}_{\leq \|f_n - f\|_p \|g\|_q \to 0} \to 0.$$

$$(7.11)$$

Hence $\sup_{x \in \mathbb{R}^d} |(f_n * g_n)(x) - (f * g)(x)| \to 0$ as $n \to \infty$.

$\square$

Lemma 7.17 (Young's inequality) *Let $p, q \in [1, \infty]$ be such that there exists $r \in [1, \infty]$ satisfying*

$$\frac{1}{r} = \frac{1}{p} + \frac{1}{q} - 1 \geq 0.$$

Let $f \in L^p$ and $g \in L^q$, then f and g are convolvable, $f * g \in L^r$ and

$$\|f * g\|_r \leq \|f\|_p \|g\|_q.$$

Note that the product of two functions f and g, belongs to some L^p space just if f is in L^p and g in L^q, with p, q conjugate indices. Young's inequality states instead that the convolution product belongs to L^p for a larger set of indices.

Proof We prove here the case $p = 1$ and $q = r \in [1, \infty[$, referring to [2, Theorem 3.4] for the proof of the other cases.

First suppose that $f \geq 0$ and $g \geq 0$ (from the proof it will be clear that this assumption is not restrictive: since we compute L^p norms of f and g, if the functions are not positive the same proof is done replacing f with $|f|$ and g with $|g|$). Define the following measure:

$$m(A) := \int_A f(y)\mathrm{d}y, \quad \forall A \subseteq \mathbb{R}^d. \tag{7.12}$$

Then

$$
\begin{aligned}
(f * g)(x) &= \int g(x - y)f(y)\mathrm{d}y = \int g(x - y)\mathrm{d}m(y) \\
&\overset{7.9}{\leq} \left(\int g(x - y)^q \mathrm{d}m(y) \right)^{\frac{1}{q}} \left(\int \mathrm{d}m(y) \right)^{1 - \frac{1}{q}} \\
&= \left(\int g(x - y)^q f(y)\mathrm{d}y \right)^{\frac{1}{q}} \left(\int f(y)\mathrm{d}y \right)^{1 - \frac{1}{q}}.
\end{aligned}
\tag{7.13}
$$

(Remark that here we use a generalization of the Hölder inequality stated in Lemma 7.9, since we use it with respect to the measure in (7.12) and not with respect to the Lebesgue measure. See for example [1, Theorem 4.6] for this generalization.)

Hence

$$
\begin{aligned}
\int ((f * g)(x))^q \mathrm{d}x &\leq \int (\int g(x - y)^q f(y)\mathrm{d}y)(\int f(y)\mathrm{d}y)^{q-1} \mathrm{d}x \\
&= \int f(y)(\int g(x - y)^q \mathrm{d}x)\mathrm{d}y (\int f(y)\mathrm{d}y)^{q-1} \\
&= \|g\|_q^q \, \|f\|_1^q,
\end{aligned}
\tag{7.14}
$$

concluding the proof with $r = q$, $p = 1$. $\qquad\qquad\square$

<u>Notation:</u> we denote by $L_c^p = L_c^p(\mathbb{R}^d)$ the set of L^p functions with compact essential support, and by L_{loc}^p the set of *measurable* functions over $\mathbb{R}^d$, such that their restriction over any compact subset K of $\mathbb{R}^d$ belongs to $L^p(K)$.

Moreover, for any $\Omega \subseteq \mathbb{R}^d$ open set, we denote by $\mathcal{C}_c^k(\Omega) := \mathcal{C}^k(\Omega) \cap \mathcal{C}_c(\Omega)$ and by $\mathcal{C}_c^\infty(\Omega) := \mathcal{C}^\infty(\Omega) \cap \mathcal{C}_c(\Omega)$.

Remark 7.18 Let $p, q \in [1, \infty]$ be conjugated indices.

1. If $f \in L_c^p$ and $g \in L_{loc}^q$, then $fg \in L^1$;
2. if $f \in L_c^p$ and $g \in L_{loc}^q$ then f and g are convolvable and $f * g$ is uniformly continuous;
3. if $f \in L_{loc}^1$ and $g \in \mathcal{C}_c(\mathbb{R}^d)$ then f and g are convolvable and $f * g$ is uniformly continuous (this is a particular case of previous point, but we underline it since it will be particularly useful);
4. if $1 \leq p \leq q \leq \infty$, then $L_{loc}^q \subseteq L_{loc}^p$.

Proposition 7.19 *Let $f \in C_c^k(\mathbb{R}^d)$ and $g \in L^1_{loc}$. Then $f * g \in C^k(\mathbb{R}^d)$ and*

$$D^\alpha(f * g) = (D^\alpha f) * g, \quad \forall |\alpha| \le k.$$

Proof We prove the result for the case $k = 1$, since the other cases for $k > 1$ follow by induction.

Fix $x \in \mathbb{R}^d$ and for every $h \in \mathbb{R}^d$ denote by

$$A_h := (f * g)(x + h) - (f * g)(x) - (\nabla f * g)(x) \cdot h,$$

where $\nabla(\cdot) = (\partial_{x_1}(\cdot), \dots, \partial_{x_d}(\cdot))$. To prove the result we show that $A_h = |h| o_h(1)$ as $|h| \to 0$, i.e.,

$$\frac{A_h}{|h|} \to 0, \quad \text{as } |h| \to 0, \tag{7.15}$$

via a direct computation.

$$
\begin{aligned}
(f * g)(x + h) - (f * g)(x) &= \int f(x + h - y)g(y)\mathrm{d}y - \int f(x - y)g(y)\mathrm{d}y \\
&= \int (f(x + h - y) - f(x - y))g(y)\mathrm{d}y \\
&= \int \left(\int_0^1 \frac{\mathrm{d}}{\mathrm{d}s} f(x + sh - y)\mathrm{d}s \right) g(y)\mathrm{d}y \\
&= \int \left(\int_0^1 \nabla f(x + sh - y)\mathrm{d}s \right) hg(y)\mathrm{d}y.
\end{aligned}
$$

$$\tag{7.16}$$

Hence

$$
\begin{aligned}
&|(f * g)(x + h) - (f * g)(x) - (\nabla f * g)(x) \cdot h| \\
&\le \int_K \left[\int_0^1 |\nabla f(x + sh - y) - \nabla f(x - y)|\mathrm{d}s \right] |h||g(y)|\mathrm{d}y,
\end{aligned}
\tag{7.17}
$$

where $K \subseteq \mathbb{R}^d$ is a compact set such that $\mathrm{supp}(f - sh - x) \subseteq K$, for $|h| \le 1$ and $s \in [0, 1]$. Finally we use the dominated convergence Theorem on the integral $\int_0^1 |\nabla f(x + sh - y) - \nabla f(x - y)|\mathrm{d}s$, which has bounded integrand, to conclude that (7.15) holds. $\qquad\square$

Corollary 7.20 *Let $f \in C_c^\infty(\mathbb{R}^d)$ and $g \in L^1_{loc}$. Then $f * g \in C^\infty(\mathbb{R}^d)$ and for all α multi-indices*

$$D^\alpha(f * g) = D^\alpha f * g.$$

If in addition $g \in L^1_c$ then $f * g \in C_c^\infty(\mathbb{R}^d)$.

Partition of unity:

In this section we construct a sequence $(\varphi_n)_n$ in $C_c^\infty(\mathbb{R}^d)$, called *mollifier* or *regularizing sequence*, which is used to regularize non-smooth functions. To this aim, we first show an example of a function ϕ in $C_c^\infty(\mathbb{R}^d)$ to prove that this space is non-empty, and then we define $(\varphi_n)_n$ and prove its regularizing properties.

Let $\psi : \mathbb{R} \to \mathbb{R}$ be the following continuous function

$$\psi(t) := \begin{cases} e^{-\frac{1}{t}} & t > 0 \\ 0 & t \leq 0. \end{cases}$$

Remark that $\psi'(t)$ is defined for $t \neq 0$ and is given by

$$\psi'(t) = \begin{cases} \frac{1}{t^2} e^{-\frac{1}{t}} & t > 0 \\ 0 & t < 0. \end{cases}$$

Putting $\psi'(0) = 0$ we have $\psi \in C^1(\mathbb{R}^d)$. Analogously for any $k \in \mathbb{N}$

$$\psi^{(k)}(t) = \begin{cases} p_k(\frac{1}{t^2}) e^{-\frac{1}{t}} & t > 0 \\ 0 & t < 0, \end{cases}$$

where $p_k(\cdot)$ is a real polynomial. Setting again $\psi^{(k)}(0) = 0$ we have that $\psi(t) \in C^k(\mathbb{R})$ for every $k \in \mathbb{N}$, hence $\psi \in C^\infty(\mathbb{R}^d)$.

Now define $h(x) := 1 - |x|^2$, which clearly belongs to $C^\infty(\mathbb{R}^d)$. Consider finally the composition $\psi \circ h : \mathbb{R}^d \to \mathbb{R}$,

$$\psi \circ h(x) = \begin{cases} e^{-\frac{1}{1-|x|^2}} & |x| < 1 \\ 0 & |x| \geq 1, \end{cases} \tag{7.18}$$

which is clearly an element of $C^\infty(\mathbb{R}^d)$ and with $\mathrm{supp}(\psi \circ h) = \overline{B(0, 1)}$, thus, $\psi \circ h \in C_c^\infty(\mathbb{R}^d)$. This provides the first example of an element in $C_c^\infty(\mathbb{R}^d)$.

Definition 7.21 Let $(\varphi_n)_n$ be a sequence in $C_c^\infty(\mathbb{R}^d)$ satisfying the following properties:

1. $\varphi_n \geq 0, \quad \forall n \in \mathbb{N}$;
2. $\int \varphi_n \mathrm{d}x = 1, \quad \forall n \in \mathbb{N}$;
3. $\mathrm{supp}\, \varphi_n \subseteq \overline{B(0, \frac{1}{n})}$.

We say that $(\varphi_n)_n$ is a regularizing function or a mollifier.

Example 7.22 Let $\varphi := \psi \circ h$ as defined in (7.18) and set

$$\varphi_1(x) = \frac{\varphi(x)}{\|\varphi\|_1},$$

which satisfies the requirements of Definition 7.21 with $n = 1$. Next for $n \in \mathbb{N}$ define

$$\varphi_n(x) = n^d \varphi_1(nx),$$

which still satisfies 7.21. The sequence $(\varphi_n)_n$ is a mollifier.

Proposition 7.23 *Let* $1 \le p < \infty$ *and take* $f \in L^p$ *and* $(\varphi_n)_n$ *a regularizing sequence. Then*

1. *$\varphi_n * f \in L^p$ and $\|\varphi_n * f\|_p \le \|f\|_p$, for all $n \in \mathbb{N}$;*
2. *$\varphi_n * f \to f$ in L^p;*
3. *$\varphi_n * f \in C^\infty(\mathbb{R}^d)$.*

Remark that the second point of the proposition is the only new result and it motivates the introduction of mollifiers for convolution products.

Proof   1. This is the application of Young's inequality 7.17 to φ_n and f;
2. To prove this second point we write

$$(\varphi_n * f)(x) - f(x) = \int f(x - y)\varphi_n(y)dy - f(x) = \int (f(x - y) - f(x))\varphi_n(y)dy.$$

Next fix $n \in \mathbb{N}$ and introduce the measure

$$m(A) := \int_A \varphi_n(y)dy.$$

Using the Hölder inequality we have:

$$|(\varphi_n * f)(x) - f(x)| \le \int |f(x - y) - f(x)|\varphi_n(y)dy$$

$$\le \left(\int |f(x - y) - f(x)|^p \varphi_n(y)dy \right)^{\frac{1}{p}} \underbrace{\left(\int \varphi_n(y)dy \right)^{1-\frac{1}{p}}}_{=1}. \quad (7.19)$$

Thus, noticing that $\varphi_n(y)^p \le \varphi_n(y)$, we have:

$$\int |(\varphi_n * f)(x) - f(x)|^p dx \le \int \int |f(x - y) - f(x)|^p \varphi_n(y)dydx$$

$$= \int \int |f(x - y) - f(x)|^p dx\,\varphi_n(y)dy$$

$$= \int \mathbb{1}|y| \le \frac{1}{n} \|\tau_y f - f\|_p^p \,\varphi_n(y)dy \qquad (7.20)$$

$$\le \sup_{|y|\le\frac{1}{n}} \|\tau_y f - f\|_p^p \underbrace{\int \varphi_n(y)dy}_{=1},$$

and $\sup_{|y|\le\frac{1}{n}}\left\|\tau_y f - f\right\|_p \to 0$ as $n \to \infty$, concluding the proof via Lemma 7.13.

3. This last point is proved by Corollary 7.20.

$\square$

Corollary 7.24 *Let* $1 \le p < \infty$ *and take* $\Omega \subseteq \mathbb{R}^d$ *open set. Then* $C_c^\infty(\Omega)$ *is dense in* $L^p(\Omega)$.

Proof Using point 3 of previous Proposition we have that $C^\infty(\mathbb{R}^d)$ is dense in $L^p(\mathbb{R}^d)$. Moreover, we already know from Proposition 7.10 that $C_c(\Omega)$ is dense in $L^p(\Omega)$. Thus, for any $\varepsilon > 0$ and any $f \in L^p(\Omega)$, there exists $\psi \in C_c(\Omega)$ such that

$$\|f - \psi\|_{L^p} \le \varepsilon.$$

Recall now that $\mathrm{dist}(\mathrm{supp}\,\psi, \Omega^c) \ge \delta > 0$ for some δ, thus, setting

$$\overline{\psi} := \begin{cases} \psi & \text{on } \Omega \\ 0 & \text{on } \Omega^c, \end{cases}$$

we have $\overline{\psi} \in C_c(\mathbb{R}^d)$. Take now a regularizing sequence $(\varphi_n)_n$ and consider

$$(\varphi_n * \psi) \in C_c^\infty(\mathbb{R}^d).$$

For any $n \ge \frac{1}{\delta}$ we have $\varphi_n * \overline{\psi} \in C_c^\infty(\Omega)$. Setting finally

$$\overline{f} = \begin{cases} f & \text{on } \Omega \\ 0 & \text{on } \Omega^c, \end{cases}$$

and remarking that

$$\left\|\varphi_n * \overline{\psi}\big|_\Omega - f\right\|_{L^p(\Omega)} = \left\|\varphi_n * \overline{\psi} - \overline{f}\right\|_{L^p(\mathbb{R}^d)}$$

we have

$$\left\|\varphi_n * \overline{\psi} - \overline{f}\right\|_{L^p(\mathbb{R}^d)} \le \underbrace{\left\|\varphi_n * \overline{\psi} - \overline{\psi}\right\|_{L^p(\mathbb{R}^d)}}_{\to 0} + \underbrace{\left\|\overline{\psi} - \overline{f}\right\|_{L^p(\mathbb{R}^d)}}_{=\|\psi-f\|_p \le \varepsilon} \le 2\varepsilon, \quad \text{for } n \text{ large enough,}$$

$$(7.21)$$

concluding the proof.

$\square$

7.3 Compactness and the Kolmogorov Criterion

In this section we prove the *Kolmogorov criterion*, which gives necessary and sufficient conditions to prove compactness of subsets of L^p spaces over $\mathbb{R}^d$.

Theorem 7.25 (Kolmogorov) *Let* $1 \leq p < \infty$ *and take* $H \subseteq L^p = L^p(\mathbb{R}^d)$. *Then* H *is relatively compact if and only if the following conditions hold:*

1. H *is bounded (i.e., there exists a positive* M *such that* $\|f\|_p \leq M$ *for all* f *in* H*);*
2. $\lim_{R \to \infty} \sup_{f \in H} \int_{|x| \geq R} |f|^p dx = 0$;
3. $\lim_{|a| \to 0} \sup_{f \in H} \|\tau_a f - f\|_p = 0$.

Proof • $\Longrightarrow$) Let H be a relatively compact (or equivalently precompact) subset of L^p and take an arbitrary $\varepsilon > 0$. Then there exists

$$\{f_j : j \in J_\varepsilon \subseteq \mathbb{N}, \ J_\varepsilon \text{ finite }\} \subseteq L^p, \quad \text{satisfying} \quad H \subseteq \bigcup_{j \in J_\varepsilon} B_{L^p}(f_j, \varepsilon).$$

Thus, H is clearly bounded. Moreover, there exists $R > 0$ such that

$$\int_{|y| \geq R} |f_j|^p dx \leq \varepsilon^p, \quad \forall j \in J_\varepsilon,$$

and by Lemma 7.13 there exists $\delta > 0$ such that

$$\left\| \tau_a f_j - f_j \right\|_p \leq \varepsilon, \quad \forall |a| \leq \delta, \ \forall j \in J_\varepsilon.$$

Thus, let $f \in H$ and choose $k \in J_\varepsilon$ such that $\|f - f_k\|_p \leq \varepsilon$. We have

$$\left(\int_{|x| \geq R} |f|^p \right)^{\frac{1}{p}} \leq \left(\int_{|x| \geq R} |f - f_k|^p \right)^{\frac{1}{p}} + \left(\int_{|x| \geq R} |f_k|^p \right)^{\frac{1}{p}} \leq 2\varepsilon.$$

Analogously, using Lemma 7.13 we get

$$\|\tau_a f - f\|_p \leq \|\tau_a f - \tau_a f_k\|_p + \|\tau_a f_k - f_k\|_p + \|f_k - f\|_p \leq 3\varepsilon, \quad \forall |a| < \delta.$$

This concludes the proof of the first implication.

• $\Longleftarrow$) Suppose now that conditions $1, 2, 3$ hold true, take $(\varphi_n)_n$ a regularizing sequence and consider the families

$$\varphi_n * H := \{\varphi_n * f : f \in H\}, \quad n \in \mathbb{N}.$$

Then, for every $f \in H$ and $n \in \mathbb{N}$ we have

(i) $\|\varphi_n * f\|_\infty \leq \|\varphi_n\|_q \|f\|_p \leq M \|\varphi_n\|_q$, for (p, q) conjugate indices.

(ii) $|(\varphi_n * f)(x) - (\varphi_n * f)(x')| \le \|\varphi_n\|_q \|\tau_x f - \tau_{x'} f\|_p$, for all $x, x' \in \mathbb{R}^d$.

(iii) $\|\varphi_n * f - f\|_p^p \le \sup_{|x| \le \frac{1}{n}} \|\tau_x f - f\|_p^p$, (the proof of this fact is contained in the proof of the second point of Proposition 7.23, thus, we omit it here).

Fix now $\varepsilon > 0$ and choose $R > 0$ and $N \in \mathbb{N}$ such that

(A) $\int_{|x| \ge R} |f|^p dx \le \varepsilon^p$, $\quad \forall f \in H$;

(B) $\|\varphi_N * f - f\|_p \le \varepsilon$, $\quad \forall f \in H$, so that, in particular $\|\varphi_N * f\|_p \le \tilde{M}$, for some positive $\tilde{M}$;

(C) item (iii) above holds for $n = N$;

thus, from conditions $(B), (C)$ and 3 the family $(\varphi_N * H)|_{B_R(0)}$ is bounded and equicontinuous in $\mathcal{C}(B_R(0))$. Then from the Ascoli-Arzelà Theorem, $(\varphi_N * H)|_{B_R(0)}$ is relatively compact in $\mathcal{C}(B_R(0))$: there exists a family

$$\{f_j \in H : \ j \in J = J_\varepsilon \subseteq \mathbb{N} \text{ finite }\} \subseteq H$$

such that $(\varphi_N * H)|_{B_R(0)} \subseteq \bigcup_{j \in J_\varepsilon} B_{\mathcal{C}(B_R(0))}\left(\varphi_N * f_j, \frac{\varepsilon}{m(B_R(0))^{\frac{1}{p}}}\right)$, where we recall that $m(\cdot)$ denotes the Lebesgue measure in $\mathbb{R}^d$. Take now f an arbitrary element of H, and $k \in J_\varepsilon$ such that

$$\|\varphi_N * f - \varphi_N * f_k\|_{\mathcal{C}(B_R(0))} \le \frac{\varepsilon}{m(B_R(0))^{\frac{1}{p}}}, \tag{7.22}$$

and write

$$\|f - f_k\|_p \le \left\|(f - f_k)\chi_{B_R(0)}\right\|_p + \left\|(f - f_k)\chi_{B_R^c(0)}\right\|_p .$$

We conclude the proof by estimating the two terms separately:

$$\left\|(f - f_k)\chi_{B_R(0)}\right\|_p \le \left\|(f - \varphi_N * f)\chi_{B_R(0)}\right\|_p + \underbrace{\left\|(\varphi_N * f - \varphi_N * f_k)\chi_{B_R(0)}\right\|_p}_{\le \varepsilon \ \text{by (7.22)}}$$

$$+ \left\|(\varphi_N * f_k - f_k)\chi_{B_R(0)}\right\|_p \le 3\varepsilon; \tag{7.23}$$

finally

$$\left\|(f - f_k)\chi_{B_R^c}\right\|_p \le \left\|f\chi_{B_R^c}\right\|_p + \left\|f_k\chi_{B_R^c}\right\|_p \le 2\varepsilon,$$

thus, $\|f - f_k\|_p \le 5\varepsilon$ and then

$$H \subseteq \bigcup_{j \in J_\varepsilon} B_{L^p}(f_k, 5\varepsilon),$$

concluding the proof. $\qquad\qquad\square$

Corollary 7.26 *Let $\Omega \subseteq \mathbb{R}^d$ be an open set and let $H \subseteq L^p(\Omega)$. We set*

$$\Omega_R := \left\{ x \in \Omega : \ \mathrm{dist}(x, \Omega^c) \geq \frac{1}{R} \right\} \cap B(0, R).$$

Then H is relatively compact in $L^p(\Omega)$ if and only if the following conditions hold true:

1. *H is bounded;*
2. *$\lim_{R \to \infty} \sup_{f \in H} \int_{\Omega \setminus \Omega_R} |f|^p \mathrm{d}x = 0$;*
3. *$\lim_{|a| \to 0} \sup_{f \in H} \|\tau_a f - f\|_{L^p(\Omega_R)} = 0$, for any $R > 0$.*

References

1. Brezis, H.E.: Functional Analysis, Sobolev Spaces and Partial Differential Equations, Universitext. Springer, New York (2010)
2. Levy, S., Hirsch, F., Lacombe, G.: Elements of Functional Analysis. Graduate Texts in Mathematics. Springer, New York (2012)

Chapter 8
Weak Topologies

8.1 How to Construct a Topology?

Given a Banach space X we denote by τ_s the topology induced by the norm, introduced and used in previous sections, and refer to it as the *strong topology*. The strong topology is the one that makes all functions in $C(X)$ continuous. A natural question arises: how do we construct the coarsest topology that makes a given collection of functions $\mathcal{F} := \{f_i, i \in I\}$ continuous? We ask this question since a coarser topology would result in a higher number of compact sets, and could for this reason be more useful in infinite dimensional contexts, which typically lack compactness.

We present in this paragraph the general procedure: let (T, τ) be a topological space and take any set S. Let $\mathcal{F} := \{f_i, i \in I\}$ a family of functions $f_i : S \to T$. We now construct the *weak topology* generated by $\mathcal{F}$ in S. We introduce:

- $\mathcal{U}_0 := \{f_i^{-1}(\omega), \ \omega \in \tau, \ i \in I\}$;
- $\mathcal{U}_1 := \{\bigcap_{i=1}^{n} \mathcal{O}_i, \ \mathcal{O}_i \in \mathcal{U}_0\}$, (i.e., finite intersections of elements in $\mathcal{U}_0$);
- $\mathcal{S} := \{\bigcup \mathcal{O}, \ \mathcal{O} \in \mathcal{U}_1\}$, (arbitrary unions of elements of $\mathcal{U}_1$).

$\mathcal{S}$ is a topology, since it contains S and is closed, by definition, under arbitrary unions and finite intersections. A basis for the topology $\mathcal{S}$ is given, by definition, by sets of the form

$$U := \bigcap_{\substack{j \in J \\ J \subseteq I, \text{finite}}} f_j^{-1}(v_j), \ v_j \in \tau, \ j \in J,$$

(in other words, $\mathcal{S}$ is the smallest topology containing $\mathcal{U}_0$). Given $x \in S$, a basis of neighbourhoods of x is given by sets of the form

$$U_x := \bigcap_{\substack{j \in J \\ J \subseteq I, \text{finite}}} f_j^{-1}(v_j), \quad \text{with } v_j \in \tau \text{ neighbourhood of } f_j(x) \text{ in } T, \ j \in J.$$

$$(8.1)$$

Consequently the following convergence property holds:

Proposition 8.1 *Let $(x_n)_n$ be a sequence in S and take $x \in S$, then $x_n \to x$ in $\mathcal{S}$ if and only if $f_j(x_n) \to f_j(x)$ for any $j \in I$.*

Proof • $\implies$) This is immediate, indeed we constructed $\mathcal{S}$ as the coarsest topology which makes all the functions f_j, $j \in I$, continuous.
- $\impliedby$) To prove this implication we must prove that, for any neighbourhood U_x of x in $\mathcal{S}$, the sequence $(x_n)_n$ is definitively contained in the neighbourhood U_x (i.e., there exists $N \in \mathbb{N}$ such that $x_n \in U_x$ for any $n > N$). We limit ourselves to neighbourhoods of the form (8.1), since they are a basis: recall that J is finite and thus, since by hypothesis $f_j(x_n) \to f_j(x)$, for all $j \in J$, there exist N_j such that $f_j(x_n) \in U_x$ for any $n > N_j$. Setting $N := \max_{j \in J} N_j$ we have proved the claim. $\quad\square$

Proposition 8.2 (Property of composition) *Using previous notation for the topological spaces (T, τ) and $(S, \mathcal{S})$, introduce a third topological space $(R, \mathcal{R})$ and let $\psi : R \to S$. Then ψ is continuous with respect to the topologies $\mathcal{R}$ and $\mathcal{S}$ if and only if $f_i \circ \psi : R \to T$ is continuous for any $i \in I$.*

Proof • $\implies$) If ψ is continuous, then clearly $f_i \circ \psi$ is a continuous function for any $i \in I$, by definition of the topology $\mathcal{S}$.
- $\impliedby$) Let now $U \in \mathcal{S}$, $U = \bigcap_{j \in J} f_j^{-1}(v_j)$, with J finite set. We have to prove that $\psi^{-1}(U) \in \mathcal{R}$.

$$\psi^{-1}(U) = \bigcap_{j \in J} \psi^{-1}(f_j^{-1}(v_j)) = \bigcap_{j \in J} (f_j \circ \psi)^{-1}(v_j).$$

Since by hypothesis $f_j \circ \psi$ is continuous for any j and v_j is an open set, then $\psi^{-1}(U)$ is open concluding the proof. $\quad\square$

8.2   Weak Topologies on Banach Spaces

We now consider a Banach space X and introduce *the weak topology on X* as the one induced by all elements in the dual X': using previous notation, $\mathcal{F} := \{f_i, i \in I\} = X'$, and thus, we consider the coarsest topology that makes all functionals in the dual space of X continuous. Precisely:

Definition 8.3 Let X be a Banach space over $\mathbb{R}$. We call weak topology in X and denote by $\sigma(X, X')$, the topology in X generated by X'.

Given a sequence $(x_n)_n$ in X we write $x_n \rightharpoonup x$ if x_n converges to x in the weak topology $\sigma(X, X')$.

Remark 8.4 Clearly we have that $\sigma(X, X') \subseteq \tau_s(X)$, where we recall that τ_s is the strong topology (or topology of the norm) in X. In general, the inclusion is strict: If a set A is weakly open then it is surely strongly open, and thus, if a set B is weakly closed it is surely strongly closed. The opposite implication in general is false, as we now show: we exhibit a sequence $(u_n)_n$ and a function u in X such that $u_n \rightharpoonup u$ but $u_n \nrightarrow u$. Let $X = L^2(\mathbb{R})$ (recall that $X' = X = L^2(\mathbb{R})$), and consider the sequence $u_n := \chi_{[n,n+1]}$. It is easy to check that such sequence is not a Cauchy sequence in X (indeed it is not a Cauchy sequence in any L^p space), and that $(u_n)_n$ is not relatively compact, since it has no subsequence which converges strongly. We now show that $(u_n)_n$ converges weakly to 0 in $L^2(\mathbb{R})$. To this aim, take $f \in X'$ and, using the Riesz Theorem 5.23, denote by $v_f \in L^2(\mathbb{R})$ the function such that

$$\langle f, w \rangle_{X',X} = (w, v_f)_{L^2(\mathbb{R})}, \quad \forall w \in L^2(\mathbb{R}).$$

Then in particular

$$
\begin{aligned}
|(u_n, v_f)_{L^2(\mathbb{R})}| = \Big| \int_{\mathbb{R}} u_n(x) v_f(x) \mathrm{d}x \Big| &= \int_n^{n+1} |v_f(x)| \mathrm{d}x \\
&\leq \Big(\int_n^{n+1} |v_f(x)|^2 \mathrm{d}x \Big)^{\frac{1}{2}} \leq \Big(\int_n^{\infty} |v_f(x)|^2 \mathrm{d}x \Big)^{\frac{1}{2}} \to 0,
\end{aligned}
\tag{8.2}
$$

hence $u_n \rightharpoonup 0$ in $L^2(\mathbb{R})$.

Remark 8.5 If X is finite dimensional, then $\sigma(X, X') = \tau_s(X)$. To show this, it is sufficient to see that any weakly converging subsequence is strongly converging (we identify from now any finite dimensional real Banach space of dimension n with $\mathbb{R}^n$). Let $x_n \rightharpoonup x \in \mathbb{R}^n$, since any $f \in X'$ has the form $\langle f, x \rangle = a \cdot x$ for some $a \in \mathbb{R}^n$, we have $a \cdot x_n \to a \cdot x$, from continuity of the scalar product. Thus, choosing $a = e_j$, where $\{e_j\}_{j=1}^n$ is the canonical base of $\mathbb{R}^n$ we obtain that $x_n^j \to x^j$ for all j, proving that $x_n \to x$ in the strong topology.

Remark 8.6 We finally prove the following important property: given $\mathcal{H}$ a real Hilbert space and $\{x_n, n \in \mathbb{N}\}$ an orthonormal system in $\mathcal{H}$, $\{x_n, n \in \mathbb{N}\}$ is a weakly closed but not strongly closed subset of $\mathcal{H}$: indeed

$$\|x_n - x_m\|^2 = \|x_n\|^2 + \|x_m\|^2 - 2(x_n, x_m) = 2, \quad \forall n \neq m.$$

Hence $(x_n)_n$ does not converge strongly in $\mathcal{H}$, but for any x in $\mathcal{H}$ we have (see Lemma 5.4)

$$\|x\| \geq \sum_{n=1}^{\infty} |(x, x_n)|^2,$$

implying that $(x, x_n) \to 0$ for all x in $\mathcal{H}$, i.e., $x_n \rightharpoonup 0$.

We now turn to proving some important results about weakly converging sequences and weak topologies.

Lemma 8.7 *Let $(x_n)_n$ be a sequence in X and let $x \in X$. If $x_n \rightharpoonup x$, then $(x_n)_n$ is bounded and*

$$\|x\| \le \liminf \|x_n\| . \tag{8.3}$$

Moreover, if $x_n \rightharpoonup x$ and $f_n \xrightarrow{X'} f$, then $\langle f_n, x_n \rangle \to \langle f, x \rangle$.

Proof Let $J : X \to X''$ be the linear isometry introduced in Definition 3.22, given by $\langle J(x), f \rangle_{X'',X'} = \langle f, x \rangle_{X',X}$, for all $x \in X$ and for all $f \in X'$. Fix $f \in X'$ and observe that $\langle J(x_n), f \rangle_{X'',X'} = \langle f, x_n \rangle_{X',X}$. Since $\langle f, x_n \rangle_{X',X} \to \langle f, x \rangle_{X',X}$ we have

$$|\langle J(x_n), f \rangle_{X'',X}| \le C_f, \quad \forall n \in \mathbb{N}.$$

We now use the Banach-Steinhaus Theorem 3.36: there exists $C \ge 0$ such that $\|x_n\|_X = \|J(x_n)\|_{X''} \le C$, for all n. Thus, $(x_n)_n$ is bounded.

Let now $f_0 \in X'$, $\|f_0\|_{X'} = 1$, $\langle f_0, x \rangle_{X',X} = \|x\|_X$ (such functional exists from Proposition 3.17). Clearly $\langle f_0, x_n \rangle \to \langle f_0, x \rangle$ and we have

$$\begin{aligned}
\|x\|_X = \langle f_0, x \rangle_{X',X} &= \lim_{n \to \infty} \langle f_0, x_n \rangle_{X',X} \le \liminf |\langle f_0, x_n \rangle| \\
&\le \liminf \|f_0\|_{X'} \|x_n\|_X = \liminf \|x_n\|_X .
\end{aligned} \tag{8.4}$$

This concludes the proof of (8.3). Finally, to prove the last part of the Lemma, we compute

$$\begin{aligned}
\langle f_n, x_n \rangle - \langle f, x \rangle &= \langle f_n, x_n \rangle - \langle f, x_n \rangle + \langle f, x_n \rangle - \langle f, x \rangle \\
&= \langle f_n - f, x_n \rangle + \underbrace{(\langle f, x_n \rangle - \langle f, x \rangle)}_{\to 0 \text{ since } x_n \rightharpoonup x},
\end{aligned} \tag{8.5}$$

and since $\|x_n\|_X$ is a bounded real sequence from previous part of the Lemma, using Schwarz inequality we get

$$|\langle f_n - f, x_n \rangle| \le \|f_n - f\|_{X'} \|x_n\|_X \xrightarrow{f_n \to f} 0, \quad \text{as } n \to \infty,$$

concluding the proof. $\qquad\square$

Lemma 8.8 *Let X be a real Banach space and let $C \subseteq X$ be a nonempty, closed and convex subset. Then C is weakly closed.*

Proof Let $x_0 \in X \setminus C$. We prove that there exists a set $A \in \sigma(X, X')$, such that $x_0 \in A$ and $A \cap C = \emptyset$. To this aim, we use the Hahn-Banach Theorem 3.12: there exists an hyperplane $\{\lambda = \alpha\}$ which separates in strict sense C and x_0, i.e.,

$$\langle \lambda, x \rangle \le \alpha - \varepsilon, \ \forall x \in C, \varepsilon > 0, \quad \text{and} \quad \langle \lambda, x_0 \rangle \ge \alpha + \varepsilon.$$

Then the set $A := \lambda^{-1}(]\alpha, \infty[) \in \sigma(X, X')$ satisfies $x_0 \in A$ and $C \cap A = \emptyset$. $\square$

Theorem 8.9 *Let X, Y be Banach spaces and let $T : X \to Y$ be a linear operator. Then*

$$T \in \mathcal{L}(X, Y) \iff T \in \mathcal{L}((X, \sigma(X, X')), (Y, \sigma(Y, Y'))).$$

Proof Preliminarily to the proof of the theorem we observe that $(X \times Y)' = X' \times Y'$.

- $\implies$) Let $T \in \mathcal{L}(X, Y)$. Using the composition property in Proposition 8.2 we have to verify that $f \circ T : (X, \sigma(X, X')) \to \mathbb{R}$ is continuous for any $f \in Y'$. But

$$(f \circ T)(x) = \langle f, Tx \rangle_{Y',Y} =: g(x).$$

Clearly g is linear and in addition

$$|g(x)| \leq |\langle f, Tx \rangle_Y| \leq \|f\|_{Y'} \|T\| \|x\|_X.$$

Hence $g \in X'$ and it is continuous.
- $\impliedby$) Suppose now that $T \in \mathcal{L}((X, \sigma(X, X')), (Y, \sigma(Y, Y')))$, take any sequence $\{x_n, y_n\}_n = \{x_n, Tx_n\}_n$ in $\Gamma(T)$ (recall Definition 3.40), such that $\{x_n, y_n\} \rightharpoonup \{x, y\}$ in $X \times Y$. Using identity $(X \times Y)' = X' \times Y'$ we have $x_n \rightharpoonup x$ in X and $Tx_n \rightharpoonup y$ in Y, thus, by the weak continuity of T we deduce that $y = Tx$. Hence $\Gamma(T)$ is weakly closed in $X \times Y$, and thus, in particular it is strongly closed. Finally using the closed graph Theorem 3.41 $T \in \mathcal{L}(X, Y)$.

$$\square$$

Theorem 8.10 (Mazur) *Let $(x_n)_n$ a sequence in X and x_0 in X be such that $x_n \rightharpoonup x_0$. Then $x_0 \in \overline{C_0}(\{x_n, \ n \in \mathbb{N}\})$ (recall Definition 3.46 of convex hull). In other words there exists a sequence of the form $y_n = \sum_{i=1}^{N_n} \lambda_j^n x_{n_j}$ with $\{x_{n_j}, \ j = 1, \ldots, N_n\} \subseteq (x_n)_n$ and $\lambda_j^n \geq 0$ such that $\sum_{j=1}^{N_n} \lambda_j^n = 1$ and $y_n \to x_0$ strongly.*

Proof Let

$$M = \overline{C_0}(\{x_n, \ n \in \mathbb{N}\}).$$

It is a closed convex subset of X. Assume by contradiction that $x_0 \notin M$. By the Hahn-Banach Theorem, (second geometric form 3.30) there exists $f \in X', f \neq 0$, and $\alpha \in \mathbb{R}$ such that the hyperplane $\{f = \alpha\}$ separates strictly M and $\{x_0\}$. Since $x_n \in M$ for all $n \in \mathbb{N}$, we have

$$\langle f, x_n \rangle \leq \langle f, x_0 \rangle - \varepsilon, \quad \forall n \in \mathbb{N}, \ \varepsilon > 0.$$

But this is a contradiction since $x_n \rightharpoonup x_0$. $\square$

Remark 8.11 A basis of neigbourhoods of $x_0 \in X$ in $\sigma(X, X')$ is given by

$$\mathcal{U}_{x_0} := \{x \in X : |\langle f_j, x - x_0 \rangle| < \varepsilon, \text{ with } f_j \in X', \ j \in J \text{ finite}, \varepsilon > 0\}.$$

Remark 8.12 If A is strongly compact, then A is weakly compact. Take indeed any arbitrary open covering of A in $\sigma(X, X')$:

$$A \subseteq \bigcup_{j \in I} \mathcal{O}_j, \quad \mathcal{O}_j \in \sigma(X, X'), \quad I \text{ any index set.}$$

Clearly $\mathcal{O}_j \in \tau_s(X)$ for all j, and by the strong compactness of A it is possible to extract a finite set $J \subseteq I$, such that $A \subseteq \bigcup_{j \in J} \mathcal{O}_j$, proving weak compactness of A.

8.3 Weak* Topology

Definition 8.13 (*Weak* topology*) Let X be a real Banach space and consider the dual space X' with the strong topology $\tau_s(X')$ induced by the norm $\|\cdot\|_{X'}$ and the weak topology $\sigma(X', X'')$. Recall the definition of the isometric linear injection $J : X \to X''$ in 3.22 and consider the closed linear subspace $J(X) \subseteq X''$. $J(X)$ is a family of functions from X' to $\mathbb{R}$, and thus, it generates a topology in X', with the same general procedure explained in Sect. 8.1. We call this topology weak* topology and we denote it by $\sigma(X', X)$.

Given a sequence $(x_n)_n$ in X we write $x_n \overset{*}{\rightharpoonup} x$ if the sequence converges to the point x in the weak* topology.

Clearly we have $\sigma(X', X) \subseteq \sigma(X', X'') \subseteq \tau_s(X')$. If X is reflexive we have $\sigma(X', X) = \sigma(X', X'')$. Moreover, a basis of neighbourhoods of $f_0 \in X'$ is given by

$$\mathcal{U}_{f_0} := \{f \in X' : \ |\langle f - f_0, x_i \rangle| < \varepsilon, \ x_i \in X, i \in J \text{ finite}, \ \varepsilon > 0\}.$$

Reasoning as we did in previous section for $\sigma(X, X')$ we have the following properties (analogous to (8.3)):

Lemma 8.14 *If $f_n \overset{*}{\rightharpoonup} f$ then $(f_n)_n$ is bounded and we have*

$$\|f\|_{X'} \leq \liminf \|f_n\|_{X'}.$$

Moreover, if $f_n \overset{*}{\rightharpoonup} f$ and $x_n \to x$ then $\langle f_n, x_n \rangle_{X', X} \to \langle f, x \rangle_{X', X}$.

Next we prove the following result, that motivates the introduction of weak* topologies:

Theorem 8.15 (Banach-Alaoglu) *Let $\overline{B}_1^{X'} = \{f \in X' : \|f\|_{X'} \leq 1\}$. Then $\overline{B}_1^{X'}$ is compact in $\sigma(X', X)$.*

Proof Let $Y = \mathbb{R}^X = \{w = (w_x)_{x \in X}, \ w_x \in \mathbb{R}\} = \prod_{x \in X} \mathbb{R}$ and, for any $x \in X$ define the projection

$$\pi_x : Y \to \mathbb{R}, \quad \pi_x(w) = w_x.$$

We call π the topology in Y generated by the collection $\{\pi_x,\ x \in X\}$. Clearly $X' \subseteq Y$ and, given $f \in X'$ we identify with w^f the element in Y that coincides with f: $w^f := (w_x^f)_{x \in X} = (\langle f, x \rangle_{X',X})_{x \in X}$. Let $\phi : X' \to Y$, be defined as $\phi(f) = w^f$. Clearly ϕ is linear and injective. We consider the map ϕ with respect to the following topologies:

$$\phi : (X', \sigma(X', X)) \to (\phi(X) \subseteq Y, \pi).$$

We claim that

1. ϕ is a linear homeomorphism;
2. $\phi(\overline{B}_1^{X'})$ is compact in (Y, π).

If such conditions hold, then

$$\overline{B}_1^{X'} = \phi^{-1}(\phi(\overline{B}_1^{X'})) \text{ is compact in } \sigma(X', X).$$

Thus, we are left to prove items 1 and 2: In order to prove continuity of ϕ we show that $\pi_x \circ \phi : (X', \sigma(X', X)) \to \mathbb{R}$ is continuous. Writing explicitly $\pi_x \circ \phi$ we have

$$X' \ni f \xrightarrow{\phi} w^f \xrightarrow{\pi_x} \pi_x(w^f) = w_x^f = \langle f, x \rangle_{X',X} \,,$$

which is a continuous function. Next to prove continuity of ϕ^{-1}, we must show that $J(x) \circ \phi^{-1} : \phi(X') \subseteq (Y, \pi) \to \mathbb{R}$ is continuous on (Y, π). But expliciting again the function $J(x) \circ \phi^{-1}$ we have: $J(x) \circ \phi^{-1}(\phi(f)) = w_x^f$, which is a continuous map with respect to π by definition. This proves the first item. Next we consider compactness of $\phi(\overline{B}_1^{X'})$:

$$\phi(\overline{B}_1^{X'}) = \{w = (w_x)_{x \in X} : w_{x+y} = w_x + w_y,\ w_{\alpha x} = \alpha w_x,\ |w_x| \le \|x\|,\ \forall x, y \in X,\ \alpha \in \mathbb{R}\}.$$

For any $x, y \in X, \alpha \in \mathbb{R}$ introduce the following sets:

$$A_{x,y} := \{w \in Y : w_{x+y} = w_x + w_y\}, \quad B_{\alpha x} := \{w \in Y : w_{\alpha x} = \alpha w_x\}$$

and $K := \prod_{x \in X}[-\|x\|, \|x\|]$, which is compact in (Y, π) thanks to the Tichonoff Theorem. Remark that $A_{x,y}$ and $B_{\alpha x}$ are closed for any $x, y \in X, \alpha \in \mathbb{R}$. Thus

$$\phi(\overline{B}_1^{X'}) = \bigcap_{x,y \in X} A_{x,y} \bigcap_{\substack{\alpha \in \mathbb{R} \\ x \in X}} B_{\alpha x} \bigcap K,$$

is a compact set. $\square$

Lemma 8.16 (Helly) *Let $f_1, \ldots, f_n \in X'$ and $\alpha = (\alpha_1, \ldots, \alpha_n) \in \mathbb{R}^n$. The following conditions are equivalent*

(i) for any $\varepsilon > 0$ there exists $x_\varepsilon \in X$, $\|x_\varepsilon\| = 1$ such that $|\langle f_i, x_\varepsilon \rangle - \alpha_i| \leq \varepsilon$, for any $i = 1, \ldots, n$;

(ii) $\left| \sum_{i=1}^n \alpha_i \beta_i \right| \leq \left\| \sum_{i=1}^n \beta_i f_i \right\|_{X'}, \quad \forall \beta \in \mathbb{R}^n.$

Proof We prove separately the two implications.

- $(i) \implies (ii)$ We write $\sum_{i=1}^n \alpha_i \beta_i = \sum_{i=1}^n \beta_i (\alpha_i - \langle f_i, x_\varepsilon \rangle) + \sum_{i=1}^n \beta_i \langle f_i, x_\varepsilon \rangle$, thus:

$$
\begin{aligned}
\left| \sum_{i=1}^n \alpha_i \beta_i \right| &\leq \sum_{i=1}^n |\beta_i| |(\alpha_i - \langle f_i, x_\varepsilon \rangle)| + |\langle \beta_i f_i, x_\varepsilon \rangle| \\
&\leq \sum_{i=1}^n |\beta_i| \varepsilon + \left\| \sum_{i=1}^n \beta_i f_i \right\|_{X'} \|x_\varepsilon\|_X \\
&\leq C\varepsilon + \left\| \sum_{i=1}^n \beta_i f_i \right\|_{X'},
\end{aligned}
\tag{8.6}
$$

  by arbitrariness of ε this implies (ii).

- $(ii) \implies (i)$ Let $\varphi : X \to \mathbb{R}^n$ given by $\varphi(x) = (\langle f_1, x \rangle, \ldots, \langle f_n, x \rangle)$. We have to prove that $\alpha \in \varphi(\overline{B}_1^X)$. Assume by contradiction that $\alpha \notin \varphi(\overline{B}_1^X)$. Since the set $\varphi(\overline{B}_1^X)$ is closed and convex, there exists $g \in (\mathbb{R}^n)'$ such that $g(y) < \gamma < g(\alpha)$, for all $y \in \varphi(\overline{B}_1^X)$. But there also exists $\beta \in \mathbb{R}^n$ such that $g(y) = \beta \cdot y$, for all $y \in \mathbb{R}^n$. Thus,

$$
\sum_{i=1}^n \beta_i \langle f_i, x \rangle < \gamma < \beta \cdot \alpha, \quad \forall \|x\| \leq 1,
$$

and exchanging x with $-x$ we have

$$
\sum_{i=1}^n \beta_i \langle f_i, -x \rangle < \gamma < \beta \cdot \alpha, \quad \forall \|x\| \leq 1,
$$

from which

$$
\left| \sum_{i=1}^n \beta_i \langle f_i, x \rangle \right| < \gamma \quad \forall \|x\| \leq 1.
$$

Thus, taking the supremum over $\{\|x\| \le 1\}$ we get

$$\left\| \sum_{i=1}^{n} \beta_i f_i \right\|_{X'} \le \gamma < \sum_{i=1}^{n} \alpha_i \beta_i,$$

which contradicts (ii). This concludes the proof. $\qquad\square$

Proposition 8.17 $J(\overline{B}_1^{X})$ *is dense in* $\overline{B}_1^{X''}$ *in the topology* $\sigma(X'', X')$.

Proof Let $\zeta \in \overline{B}_1^{X''}$ and let V be a neighbourhood of ζ in the $\sigma(X'', X')$ topology:

$$V = \{\eta \in \overline{B}_1^{X''} : |\langle \eta - \zeta, f_i \rangle_{X'',X'}| < \varepsilon, \ f_i \in X', \ i = 1, \ldots, n, \ \varepsilon > 0\}.$$

We look for $x \in \overline{B}_1^{X}$ such that $J(x) \in V$. This means that $J(x)$ must satisfy equivalently:

$$|\langle J(x), f_i \rangle_{X'',X'} - \langle \zeta, f_i \rangle_{X'',X'}| < \varepsilon, \ \text{ or } \ |\langle f_i, x \rangle_{X',X} - \langle \zeta, f_i \rangle_{X'',X'}| < \varepsilon, \quad \forall i = 1, \ldots, n. \tag{8.7}$$

Set $\alpha_i = \langle \zeta, f_i \rangle$, for every $i = 1, \ldots, n$ and observe that, for every $\beta = (\beta_1, \ldots, \beta_n) \in \mathbb{R}^n$ we have

$$\sum_{i=1}^{n} \alpha_i \beta_i = \sum \beta_i \langle \zeta, f_i \rangle_{X'',X'} = \left\langle \zeta, \sum_{i=1}^{n} \beta_i f_i \right\rangle_{X'',X'},$$

from which we obtain that

$$\left| \sum_{i=1}^{n} \alpha_i \beta_i \right| \le \|\zeta\|_{X''} \left\| \sum \beta_i f_i \right\|_{X'} \le \left\| \sum \beta_i f_i \right\|_{X'}.$$

Hence, applying Helly's Lemma we obtain $x_\varepsilon \in \overline{B}_1^{X}$ such that condition (8.7) holds with $x = x_\varepsilon$ for all $i = 1, \ldots, n$ as claimed. $\qquad\square$

8.4 Consequences on Reflexivity

Theorem 8.18 (Kakutani) X *is reflexive if and only if* $\overline{B}_1^{X}$ *is compact in* $\sigma(X, X')$.

Proof We prove separately the two implications.

- $\implies$) Suppose that X is reflexive, i.e., $J(X) = X''$. In particular we have $J(\overline{B}_1^{X}) = \overline{B}_1^{X''}$ and $\overline{B}_1^{X} = J^{-1}(\overline{B}_1^{X''})$. We know from the Banach-Alaoglu Theorem 8.15 that $\overline{B}_1^{X''}$ is compact in $\sigma(X'', X')$. Thus, we consider the function

$$J^{-1} : (X'', \sigma(X'', X')) \to (X, \sigma(X, X')),$$

which is continuous: indeed let $f \in X'$, then $f \circ J^{-1} : (X'', \sigma(X'', X')) \to \mathbb{R}$ can be written as $f \circ J^{-1}(\zeta) = \langle \zeta, f \rangle_{X'',X'}$, which is continuous on $(X'', \sigma(X'', X'))$ by definition. Now recall that $\overline{B}_1^X = J^{-1}(\overline{B}_1^{X''})$, which is hence compact in $\sigma(X', X)$, since $\overline{B}_1^{X''}$ is compact in $\sigma(X'', X')$ and J^{-1} is continuous with respect to these topologies.

- $\Longleftarrow$) Suppose now that $\overline{B}_1^X$ is compact in $\sigma(X, X')$. Consider $J(\overline{B}_1^X)$, which, thanks to previous Proposition is dense in $\overline{B}_1^{X''}$ with respect to $\sigma(X'', X')$. The linear map J is continuous with respect to the strong topologies $\tau_s(X) \to \tau_s(X'')$, thus, from Theorem 8.9 it is also continuous with respect to $\sigma(X, X') \to \sigma(X'', X''')$, and since $\sigma(X'', X') \subseteq \sigma(X'', X''')$, J is also continuous as map from $\sigma(X, X')$ to $\sigma(X'', X')$. Hence $J(\overline{B}_1^X)$ is closed in $(X'', \sigma(X'', X'))$, being continuous image of a compact set. Finally, using that it is also dense in $\overline{B}_1^{X''}$ we have $J(\overline{B}_1^X) = \overline{B}_1^{X''}$. Now let $\zeta \in X', \zeta \neq 0$, and define $\overline{x}$ as the element of $\overline{B}_1^X$ such that $\frac{\zeta}{\|\zeta\|_{X''}} = J(\overline{x})$. Then

$$\zeta = \|\zeta\|_{X''} J(\overline{x}) = J(\|\zeta\|_X \overline{x}),$$

implying that $X'' = J(X)$.

$\square$

Lemma 8.19 *Let X, Y be real Banach spaces and let $(T_n)_n$ be a bounded sequence in $\mathcal{L}(X, Y)$ satisfying the following property: there exists $S \subseteq X$, dense in X such that the sequence $(T_n s)_n$ converges in Y for every $s \in S$. Then there exists $T \in \mathcal{L}(X, Y)$ such that $T_n x \to T x$ for all $x \in X$.*

Proof Let $x \in X$ and let $\varepsilon > 0$. Select $x_0 \in S$ such that $\|x - x_0\| < \varepsilon$. We have

$$\begin{aligned} \|T_n x - T_m x\| &\leq \|T_n x - T_n x_0\| + \|T_n x_0 - T_m x_0\| + \|T_m x_0 - T_m x\| \\ &\leq (\|T_m\| + \|T_n\|) \|x - x_0\| + \|T_n x_0 - T_m x_0\|. \end{aligned} \tag{8.8}$$

Since $\|T_n\| \leq M$ for every $n \in \mathbb{N}$ we have that

$$\|T_n x - T_m x\| \leq 2M\varepsilon + \|T_n x_0 - T_m x_0\|.$$

Thus, $(T_n x)_n$ is a Cauchy sequence in Y and we have $T_n x \to y \in Y$. Setting $y = T x$ we see immediately that T is linear and that $\|T x\| \leq \lim_{n \to \infty} \|T_n x\| \leq M \|x\|$. Hence $T \in \mathcal{L}(X, Y)$. $\square$

Proposition 8.20 *Let X be a separable Banach space and let $(f_n)_n$ be a bounded sequence in X'. Then there exists $f \in X'$ and $(f_{n_k})_k$ subsequence of $(f_n)_n$ such that*

$$\langle f_{n_k}, x \rangle_{X',X} \to \langle f, x \rangle_{X',X} \quad \forall x \in X,$$

i.e., $f_{n_k} \overset{*}{\rightharpoonup} f$.

Proof Let $S \subseteq X$ be a countable and dense subset of X. For any $y \in S$ consider the real sequence $(\langle f_n, y \rangle)_n$, which is bounded since $(f_n)_n$ is, and thus, is a relatively compact sequence in $\mathbb{R}$. Applying diagonal procedure we obtain a subsequence $(f_{n_k})_k$ such that $(\langle f_{n_k}, y \rangle_{X',X})_n$ converges for any $y \in S$. Applying previous lemma we get weak star convergence. $\qquad\square$

Proposition 8.21 *Let X be a reflexive Banach space and let $M \subseteq X$ be a strongly closed subspace. Then M is reflexive.*

Proof In the space $(M, \|\cdot\|_M)$, where $\|\cdot\|_M$ is the norm in M induced by $\|\cdot\|_X$, a priori two weak topologies are defined: $\sigma(M, M')$ and $\sigma(X, X') \cap M$. Actually these two topologies coincide (this is a consequence of the Hahn-Banach Theorem). Hence using the Kakutani Theorem 8.18 we see that $\overline{B}_1^X$ is compact in $\sigma(X, X')$, and then $\overline{B}_1^M = \overline{B}_1^X \cap M$ is compact in $\sigma(M, M') = \sigma(X, X')|_M$. $\qquad\square$

Proposition 8.22 *Let X be a real Banach space. Then X is reflexive if and only if X' is reflexive.*

Proof We prove separately the two implications.

- $\Longrightarrow$) Suppose that X is reflexive. By the Banach-Alaouglu Theorem $\overline{B}_1^{X'}$ is compact in $\sigma(X', X) = \sigma(X', X'')$. Hence X' is reflexive by the Kakutani Theorem.
- $\Longleftarrow$) Suppose X' reflexive. From previous implication X'' is reflexive, and thus, also $J(X)$ is (since $J(X)$ is a strongly closed subspace of X'', reflexivity follows from Proposition 8.21). Moreover, $\overline{B}_1^X = J^{-1}(\overline{B}_1^{J(X)})$, and we notice that $J^{-1} : J(X) \to X$ is linear and $\tau_s(X'') \to \tau_s(X)$ continuous. Hence J^{-1} is $\sigma(X', X'') \to \sigma(X, X')$ continuous. By reflexivity of $J(X)$ and the Kakutani Theorem $\overline{B}_1^{J(X)}$ is $\sigma(X'', X''')$ compact, hence $\overline{B}_1^X$ is compact in $\sigma(X, X')$, being the continuous image of a compact set. By the Kakutani Theorem we conclude that X is reflexive.

$\qquad\square$

Corollary 8.23 *X is reflexive and separable if and only if X' is reflexive and separable.*

Proof We just need to show the first implication, since the latter has already been proved. Suppose that X is reflexive and separable, then $J(X) = X''$ is reflexive and separable. Thus, using the other implication we conclude that X' is reflexive and separable. $\qquad\square$

Proposition 8.24 *Let X be a reflexive real Banach space and let $(x_n)_n$ be a bounded sequence in X. Then there exists a subsequence $(x_{n_k})_k$ and $x \in X$ such that $x_{n_k} \rightharpoonup x$.*

Proof Suppose first that X is separable and consider the sequence $\zeta_n := J(x_n)$, which is bounded in the reflexive and separable Banach space X''. We already know that there exists $(\zeta_{n_k})_k$ such that $\zeta_{n_k} \overset{*}{\rightharpoonup} \zeta$ in X'', i.e.

$$\langle J(x_{n_k}), f \rangle_{X'',X'} \to \langle J(x), f \rangle_{X',X}, \quad \forall f \in X',$$

or equivalently

$$\langle f, x_{n_k} \rangle_{X',X} \to \langle f, x \rangle_{X',X}, \quad \forall f \in X', \quad \text{or} \quad x_{n_k} \rightharpoonup x.$$

Suppose now that X is not separable and define

$$M := \overline{\mathrm{lsp}(\{x_n, \ n \in \mathbb{N}\})},$$

which is a closed subspace of X and thus, is reflexive and separable. By previous step there exists $(x_{n_k})_k$ such that $x_{n_k} \rightharpoonup x$ in $\sigma(M, M')$, but we have seen that $\sigma(M, M') = \sigma(X, X')|_M$, thus, this concludes the proof. $\qquad\square$

8.5 Uniform Convexity and the Milman Theorem

Definition 8.25 (Uniform convexity) We say that a Banach space X is uniformly convex if for any $\varepsilon > 0$ there exists $\delta_\varepsilon > 0$ such that, for every $x, y \in \overline{B}_1^X$ such that $\|x - y\| \geq \varepsilon$,

$$\left\| \frac{x + y}{2} \right\| \leq 1 - \delta_\varepsilon.$$

The first example we give, and the one to have in mind is the following: $\mathbb{R}^d$ endowed with the euclidean norm is uniformly convex. If instead we consider the norms $\|\cdot\|_1$ and $\|\cdot\|_\infty$, defined as

$$\|x\|_1 := \sum_{i=1}^{d} |x_i|, \quad \|x\|_\infty := \max_i |x_i|,$$

then $\mathbb{R}^d$ is not uniformly convex with respect to any of these norms. We leave the proof of this fact as an exercise and we now prove the following relevant case:

Proposition 8.26 *Any Hilbert space $\mathcal{H}$ is uniformly convex.*

Proof The proof is a consequence of parallelogram identity, that we now recall: for any x, y in $\mathcal{H}$

$$\|x + y\|^2 + \|x - y\|^2 = 2(\|x\|^2 + \|y\|^2).$$

If $\|x\|$, $\|y\| \le 1$, we have $\|x + y\|^2 \le 4 - \|x - y\|^2 = 4\left(1 - \left\|\frac{x-y}{2}\right\|^2\right)$. Thus, if $\|x - y\| \ge \varepsilon$ we obtain

$$\frac{1}{4}\|x + y\|^2 = \left\|\frac{x + y}{2}\right\|^2 \le 1 - \frac{\varepsilon^2}{4},$$

thus

$$\left\|\frac{x + y}{2}\right\| \le \left(1 - \frac{\varepsilon^2}{4}\right)^{\frac{1}{2}} = 1 - \underbrace{\left[1 - \left(1 - \frac{\varepsilon^2}{4}\right)^{\frac{1}{2}}\right]}_{=\delta_\varepsilon},$$

concluding the proof. $\qquad\square$

Theorem 8.27 (Milman) *Any uniformly convex Banach space is reflexive.*

Proof First we show that

$$J(\overline{B}_1^X) = \overline{B}_1^{X''}. \tag{8.9}$$

If Eq. (8.9) holds, then the theorem can be proved as follows: take $\zeta \in X''$, $\zeta \ne 0$ and consider the corresponding element $x \in \overline{B}_1^X$ such that $J(x) = \frac{\zeta}{\|\zeta\|}$. Then $\zeta = J(\|\zeta\| x)$, and this gives $J(X) = X''$ concluding the proof. Thus, we are left to show Eq. (8.9). Take a sequence $(\zeta_n)_n$, $\zeta_n = J(x_n)$ in $J(\overline{B}_1^X)$, with $\|x_n\| = 1$ and such that

$$\zeta_n \xrightarrow{X''} \zeta.$$

We have:

$$\|x_n - x_m\|_X = \|\zeta_n - \zeta_m\|_{X''} \to 0, \quad \text{as } n, m \to \infty.$$

Hence $x_n \to x$ and $\zeta_n = J(x_n) \to J(x)$, proving that $J(\overline{B}_1^X)$ is closed in $\overline{B}_1^{X''}$. We conclude the proof showing that $J(\overline{B}_1^X)$ is dense in $\overline{B}_1^{X''}$. In particular we prove that, given any $\zeta \in \overline{B}_1^{X''}$ and any $\varepsilon > 0$, there exists $x \in \overline{B}_1^X$ such that $\|J(x) - \zeta\|_{X''} \le \varepsilon$.

Let $\delta = \delta_\varepsilon > 0$ be the modulus of uniform convexity associated with ε and select $f \in X'$, $\|f\|_{X'} = 1$ such that

$$\langle \zeta, f \rangle_{X'',X'} > 1 - \frac{\delta}{2}.$$

Then define

$$V := \left\{ \eta \in X'' : |\langle \eta - \zeta, f \rangle_{X'',X'}| \le \frac{\delta}{2} \right\}.$$

$\mathcal{V}$ is a neigbourhood of ζ in $\sigma(X'', X')$ and, by density of $J(\overline{B}_1^X)$ in $\overline{B}_1^{X''}$ with respect to the topology $\sigma(X'', X')$, there exists $x \in \overline{B}_1^X$ such that $J(x) \in \mathcal{V}$, i.e.,

$$|\langle J(x) - \zeta, f \rangle_{X'',X'}| \leq \frac{\delta}{2}, \quad \text{or equivalently } |\langle f, x \rangle_{X',X} - \langle \zeta, f \rangle_{X'',X'}| \leq \frac{\delta}{2}.$$

Thus, we have found $x \in \overline{B}_1^X$ satisfying density property in $\sigma(X'', X')$. We conclude claiming that the same point x is also sufficiently close to ζ in the strong topology $\tau_s(X'')$, i.e., $\|J(x) - \zeta\|_{X''} \leq \varepsilon$.

Suppose indeed by contradiction that $\|J(x) - \zeta\|_{X''} > \varepsilon$. This means that

$$\zeta \in (J(x) + \varepsilon \overline{B}_1^{X''})^c =: \mathcal{W}.$$

By the Banach-Alaoglu Theorem the set $J(x) + \varepsilon \overline{B}_1^{X''}$ is closed in $\sigma(X'', X')$ and then $\mathcal{W}$ is open. Hence we have

(i) $\zeta \in \mathcal{V} \cap \mathcal{W}$, open set in $\sigma(X'', X')$
(ii) $\|\zeta\|_{X''} = 1$.

Thus, by density of $J(\overline{B}_1^X)$ in $\overline{B}_1^{X''}$ in $\sigma(X'', X')$ there exists $y \in \overline{B}_1^X$ such that

$$J(y) \in \mathcal{V} \cap \mathcal{W}.$$

Thus, we have that

$$|\langle f, y \rangle_{X',X} - \langle \zeta, f \rangle_{X'',X'}| \leq \frac{\delta}{2}, \;\; |\langle f, x \rangle_{X',X} - \langle \zeta, f \rangle_{X'',X'}| \leq \frac{\delta}{2}, \text{for any} J(x), J(y) \in \mathcal{V}.$$

Thus, rearranging and summing we obtain

$$2 \langle \zeta, f \rangle_{X'',X'} - \delta < \langle f, x + y \rangle_{X',X} \leq \|f\|_{X'} \|x + y\| = \|x + y\|.$$

Now, recalling that $\langle \zeta, f \rangle_{X'',X'} > 1 - \frac{\delta}{2}$ we obtain

$$\|x + y\| > 2 \langle \zeta, f \rangle_{X'',X'} - \delta > 2 - 2\delta, \quad \text{or equivalently} \quad \left\| \frac{x + y}{2} \right\| > 1 - \delta.$$

Since $x, y \in \overline{B}_1^X$, this implies $\|x - y\| \leq \varepsilon$, which in turn gives $\|J(y) - J(x)\|_{X''} = \|x - y\| \leq \varepsilon$. But by construction $J(y) \in \mathcal{W}$ and, recalling the definition of $\mathcal{W}$, this means that $\|J(y) - J(x)\|_{X''} > \varepsilon$ giving a contradiction. $\square$

Proposition 8.28 *Let X be a uniformly convex Banach space and let $(x_n)_n$ be a sequence in X such that*

(i) $x_n \rightharpoonup x$;
(ii) $\|x\| \geq \limsup \|x_n\|$.

Then $x_n \to x$ strongly.

Proof Since $\|x\| \le \liminf \|x_n\|$, we actually have $\|x_n\| \to \|x\|$. We suppose that $\|x\| > 0$ and also that $\|x_n\| > 0$ for all $n \in \mathbb{N}$. Define

$$y_n := \frac{x_n}{\|x_n\|}, \quad y = \frac{x}{\|x\|}.$$

Clearly $y_n \rightharpoonup y$ and $y_n + y \rightharpoonup 2y$; Moreover, $\|y_n\| = 1$ for all n and also $\|y\| = 1$. Assume by contradiction that $(y_n)_n$ does not converge strongly to y. This implies that there exists $\varepsilon > 0$ and a subsequence $(y_{n_k})_k$ such that $\|y - y_{n_k}\| \ge \varepsilon$, for all $k \in \mathbb{N}$. Then, calling $\delta = \delta_\varepsilon$ the modulus of uniform convexity of X associated with ε, we obtain

$$\|y_{n_k} + y\| \le 2 - 2\delta.$$

But by weak convergence of y_n to y we have that

$$2 = \|2y\| \le \liminf \|y_{n_k} + y\| \le 2 - 2\delta,$$

obtaining a contradiction that concludes the proof. $\qquad\square$

8.6 The Dual of L^p

In this last section we go back to L^p spaces and characterize, when possible, their dual spaces. First we use some results of the previous section related to uniform convexity in order to prove that any L^p space with $1 < p < \infty$ is reflexive (Corollary 8.31). This will be crucial to define the dual spaces.

We preliminarily state the first and second Clarkson inequalities:

Lemma 8.29 – *Let $2 \le p < \infty$ and let $E \subseteq \mathbb{R}^d$ be a measurable set. Then for every $f, g \in L^p(E)$ we have*

$$\left\|\frac{f + g}{2}\right\|_p^p + \left\|\frac{f - g}{2}\right\|_p^p \le \frac{1}{2}(\|f\|_p^p + \|g\|_p^p).$$

– *Let $1 < p < 2$ and let $E \subseteq \mathbb{R}^d$ measurable, denote by q the conjugate index to p. Then for every $f, g \in L^p(E)$ we have*

$$\left\|\frac{f + g}{2}\right\|_p^q + \left\|\frac{f - g}{2}\right\|_p^q \le \left(\frac{1}{2}(\|f\|_p^p + \|g\|_p^p)\right)^{\frac{q}{p}}.$$

We refer to [1, Theorem 4.10, Step 1], for a proof. A first direct consequence is the following Proposition.

Proposition 8.30 *Let $E \subseteq \mathbb{R}^d$ be any measurable set and let $1 < p < \infty$. Then $L^p(E)$ is uniformly convex.*

Proof We consider first the case $2 \le p < \infty$ and then $1 < p < 2$.

- $\underline{2 \le p < \infty}$ Let $f, g \in L^p(E)$ with $\|f\|_p, \|g\|_p \le 1$. Then we have:

$$\left\| \frac{f+g}{2} \right\|_p^p \le \frac{1}{2}(\|f\|^p + \|g\|^p) - \left\| \frac{f-g}{2} \right\|_p^p \le 1 - \left\| \frac{f-g}{2} \right\|_p^p.$$

If $\|f - g\|_p \ge \varepsilon$ we obtain

$$\left\| \frac{f+g}{2} \right\|_p \le \left(1 - \frac{\varepsilon^p}{2^p}\right)^{\frac{1}{p}} = 1 - \left[1 - \left(1 - \frac{\varepsilon^p}{2^p}\right)^{\frac{1}{p}}\right] = 1 - \delta,$$

with $\delta = \delta_\varepsilon = 1 - \left(1 - \frac{\varepsilon^p}{2^p}\right)^{\frac{1}{p}} > 0$.

- $\underline{1 < p < 2}$ Let f, g as above. Then $\left\| \frac{f+g}{2} \right\|_p^p \le 1 - \left\| \frac{f-g}{2} \right\|_p^q \le 1 - \frac{\varepsilon^q}{2^q}$. So that, with a computation similar to the one performed in previous case, we get

$$\left\| \frac{f+g}{2} \right\|_p \le \left(1 - \frac{\varepsilon^q}{2^q}\right)^{\frac{1}{q}} = 1 - \delta_\varepsilon > 0.$$

$\square$

Corollary 8.31 *Let $E \subseteq \mathbb{R}^d$ measurable and let $1 < p < \infty$, then $L^p(E)$ is reflexive.*

We now enter into the discussion about duals of L^p spaces. Let $E \subseteq \mathbb{R}^d$ be a measurable set and $p, q \in [1, \infty]$ be conjugate indices. Given $v \in L^q(E)$ we define the map $L_v : L^p(E) \to \mathbb{R}$ as

$$L_v(u) = \int_E v(x)u(x)dx, \quad u \in L^p(E).$$

First of all we remark that L_v is well defined, since

$$|L_v(u)| \le \int_E |vu|dx \le \|v\|_q \|u\|_p, \quad \forall u \in L^p(E).$$

Moreover, thanks to the linearity of the integral, the map L_v is linear. Thus, $L_v \in (L^p(E))'$ and $\|L_v\|_{(L^p)'} \le \|v\|_q$.

Lemma 8.32 $\|L_v\|_{(L^p)'} = \|v\|_q$.

Proof Suppose that $v \ne 0$. We prove the Lemma exhibiting a function $u \in L^p(E)$, $\|u\|_p = 1$ and such that $|L_v(u)| \ge \|v\|_q$.

- $\underline{(1 < p < \infty)}$ We set

$$u(x) := \begin{cases} |v(x)|^{q-2}v(x) & v(x) \neq 0 \\ 0 & v(x) = 0. \end{cases}$$

Indeed $u \in L^p(E)$ since $|u(x)|^p = |v(x)|^q$, (where we used that, being conjugated indices, p and q satisfy $p = \frac{q}{q-1}$). Moreover,

$$\|u\|_p = (\|v\|_q^q)^{\frac{1}{p}} = (\|v\|_q^q)^{1-\frac{1}{q}} = \|v\|_q^q \|v\|_q^{-1}.$$

Hence $\|u\|_p \|v\|_q = \|v\|_q^q$. Finally computing

$$L_v(u) = \int_E |v|^q dx = \|v\|_q^q = \|u\|_p \|v\|_q ,$$

we conclude that

$$L_v\left(\frac{u}{\|u\|_p}\right) = \|v\|_q .$$

- $\underline{(p = \infty, q = 1)}$ We set

$$u(x) := \begin{cases} \frac{v(x)}{|v(x)|} & v(x) \neq 0 \\ 0 & v(x) = 0, \end{cases}$$

and we notice that $|u(x)| \leq 1$ for almost every $x \in E$, and $|u(x)| = 1$ on the set in which $v(x) \neq 0$. Hence $u \in L^\infty$ and $\|u\|_\infty = 1$. We have

$$L_v(u) = \int_E \frac{v(x)}{|v(x)|} v(x) dx = \int_E |v(x)| dx = \|v\|_1 ,$$

concluding the proof of this case too.
- $\underline{(p = 1, q = \infty)}$ Fix any real, positive ε and a corresponding set $A = A_\varepsilon \subseteq E$, measurable and such that $0 < m(A) < \infty$ and $|v(x)| \geq \|v\| - \varepsilon$, for almost every $x \in A$. Then set

$$u(x) := \begin{cases} \frac{v(x)}{|v(x)|} & \text{for almost every } x \in A \\ 0 & \text{for almost every } x \in E \setminus A. \end{cases}$$

Then $|u| = \chi_A$, so that, since $\int_E |u| dx = m(A)$, $u \in L^1(E)$ and $\|u\|_1 = m(A)$. Compute now

$$L_v(u) = \int_A \frac{v(x)}{|v(x)|} v(x)\mathrm{d}x = \int_A |v(x)|\mathrm{d}x \geq \int_A (\|v\|_\infty - \varepsilon)\mathrm{d}x \tag{8.10}$$
$$= (\|v\|_\infty - \varepsilon)m(A) = (\|v\|_\infty - \varepsilon)\|u\|_1 .$$

Thus, $L_v\left(\frac{u}{\|u\|_1}\right) \geq \|v\|_\infty - \varepsilon$ and from arbitrariness of ε this implies $\|L_v\|_{(L^1(E))'} \geq \|v\|_\infty$.

$\square$

<u>Exercise:</u> Let $E \subseteq \mathbb{R}^d$ be a measurable set of finite measure. Then for every $v \in L^\infty(E)$ we have

$$\|v\|_{L^p(E)} \to \|v\|_\infty, \quad \text{as } p \to \infty.$$

Next we prove the following theorem:

Theorem 8.33 *Let $E \subseteq \mathbb{R}^d$ be any measurable set, let $1 < p < \infty$ and $q = p'$ the conjugate index. Let $T : L^q(E) \to L^p(E)'$ be given by*

$$\langle Tu, f \rangle_{(L^p)', L^p} = \int uf\mathrm{d}x, \quad \forall f \in L^p.$$

Then T is linear, continuous, isometric, injective and surjective.

We need these two preliminary lemmas.

Lemma 8.34 *Let $E \subseteq \mathbb{R}^d$ be measurable, take $1 \leq p \leq \infty$ and let q be its conjugate index: $\frac{1}{p} + \frac{1}{q} = 1$. Suppose that $f \in L^p(E)$ is such that $\int_E f(x)g(x)\mathrm{d}x = 0$ for all $g(x) \in L^q(E)$. Then $f(x) = 0$ for almost every $x \in E$.*

Proof Let $1 < p < \infty$ and define

$$g(x) := \begin{cases} |f(x)|^{p-2}f(x) & x : f(x) \neq 0 \\ 0 & x : f(x) = 0. \end{cases}$$

We have already seen that $|g(x)|^q = |f(x)|^p$, implying that $g \in L^q(E)$. Hence we have

$$0 = \int_E f(x)g(x)\mathrm{d}x = \int_E |f(x)|^p\mathrm{d}x = 0,$$

since $fg = |f|^p$.

If instead $p = 1$, define

$$g(x) := \begin{cases} \frac{f(x)}{|f(x)|} & x : f(x) \neq 0 \\ 0 & x : f(x) = 0 \end{cases}$$

and notice that $g \in L^\infty(E)$ and that $\int_E f(x)g(x)\mathrm{d}x = \int_E |f(x)|\mathrm{d}x$, concluding the proof.

Finally if $p = \infty$ and $q = 1$, given any $R > 0$ define $g_R(x) = f(x)\chi_{B(0,R)\cap E}(x)$. Clearly $g \in L^1(E)$ and we have

$$0 = \int f(x)g_R(x)\mathrm{d}x = \int_{E \cap B(0,R)} |f(x)|^2\mathrm{d}x.$$

Hence $f(x) = 0$ for almost every $x \in E \cap B(0, R)$ and $\forall R > 0$. From arbitrariness of $R > 0$, we conclude that $f(x) = 0$ almost everywhere. $\qquad\square$

Lemma 8.35 *Let X be a Banach space and let $Y \subseteq X$ a subspace. Then Y is dense in X if and only if the following holds:*

$$\textit{if } h \in X' \quad \textit{and} \quad h|_Y = 0, \textit{ then } h = 0.$$

Proof We prove separately the two implications.

- $\Longrightarrow$) This first implication is immediate since a continuous function which vanishes on a dense set is necessarily the zero function.
- $\Longleftarrow$) Assume by contradiction that Y is not dense in X. Then there exists $z \in X \setminus Y$ such that $\mathrm{dist}(z, Y) = \delta > 0$. Let $h \in X'$ be such that $h|_Y = 0$, call $g := h|_Y$ and observe that $g \in Y'$. Thus, by the Hahn-Banach Theorem there exists a functional $G \in X'$ such that $G|_Y = g = 0$ and $G(z) = \delta$. But this contradicts the fact that $G = 0$.

$\qquad\square$

Proof (*Proof of Theorem* 8.33) Linearity, continuity and isometry have already been proved at the beginning of this section.

Injectivity: let $u, v \in L^q$ and let $Tu = Tv$, i.e., $\langle Tu, f\rangle = \langle Tv, f\rangle$, for all $f \in L^p$, so that $\langle T(u - v), f\rangle = \int_E (u - v)f\,\mathrm{d}x$, for all $f \in L^p(E)$. Thus, $u = v$ in $L^q(E)$.

Surjectivity: We know that $T \in \mathcal{L}(L^q, (L^p)')$ and $\|T\| = 1$. In addition $T(L^q)$ is closed in $(L^p)'$. Indeed let $(g_n)_n = (Tf_n)_n$, $g_n \to g$ in $(L^p)'$. Hence $\|f_n - f_m\|_q = \|g_n - g_m\|_{p'}$ and $(f_n)_n$ is a Cauchy sequence in L^q. Hence there exists $f \in L^q$ such that $f_n \to f$ and $Tf_n \to Tf$. In order to prove that T is surjective, i.e., $T(L^q) = (L^p)'$ we prove that $T(L^q)$ is dense in $(L^p)'$. Let $h \in (L^p)''$ be such that $h|_{T(L^q)} = 0$. We claim that $h = 0$. Since $\langle h, Tu\rangle_{(L^p)'',(L^p)'} = 0$ for all $u \in L^q$ and using reflexivity of L^p we write $h = J(\tilde{h})$, with $\tilde{h} \in L^p$. Hence we have, for every $u \in L^q$:

$$\langle h, Tu\rangle_{(L^p)'',(L^p)'} = \Big\langle J(\tilde{h}), Tu\Big\rangle_{(L^p)'',(L^p)'} = \Big\langle Tu, \tilde{h}\Big\rangle_{(L^p)'.L^p} = \int_E u\tilde{h}\mathrm{d}x = 0, \quad \forall u \in L^q.$$

Hence $\tilde{h} = 0$ by Lemma 8.34 and $h = J(\tilde{h}) = 0$ concluding the proof. $\qquad\square$

This theorem proves that $(L^p)' \simeq L^q$ for any $1 < p < \infty$ with q conjugate index to p. The same conclusion holds true if $p = 1$ and $q = \infty$:

Table 8.1 Dual of L^p spaces

L^p Space	Separable	Reflexive	Dual space
$p = 1$	Yes	No	L^∞
$1 < p < \infty$	Yes	Yes	$L^{p'}$
$p = \infty$	No	No	strictly contains L^1

Proposition 8.36 $(L^1)' \simeq L^\infty$.

We refer to [1, Theorem 4.14] for a proof. We conclude instead the discussion on dual of L^p spaces remarking that it is not true instead that $(L^\infty)' = L^1$. We now prove indeed that, setting

$$T : L^1(\mathbb{R}^d) \to L^\infty(\mathbb{R}^d)', \quad \langle Tu, f \rangle_{(L^\infty)', L^\infty} = \int_E uf\,dx, \quad \text{for } f \in L^\infty(\mathbb{R}^d),$$

we have $L^1 \subsetneq (L^\infty)'$.

To this aim, let $Y = \mathcal{C}_0(\mathbb{R}^d)$, which is a closed subspace of $L^\infty(\mathbb{R}^d)$ and define $Au = u(0)$, for all $u \in Y$, so that $A \in Y'$ and $\|A\|_{Y'} = 1$. By the Hahn-Banach Theorem there exists $\tilde{A} \in (L^\infty)'$ such that $\left\|\tilde{A}\right\|_{(L^\infty)'} = 1$ and $\tilde{A}|_Y = T$. This means that $\tilde{A}u = u(0)$ for all $u \in Y$. Assume now by contradiction that T is surjective: $T(L^1(E)) = L^\infty(E)'$. Then in particular there exists $w \in L^1(E)$ such that $\left\langle \tilde{A}, u \right\rangle_{(L^\infty)', L^\infty} = \int wu\,dx$ for all $u \in L^\infty$. Thus, restricting to Y we have $\left\langle \tilde{A}, u \right\rangle = u(0)$ for all $u \in Y$. But taking any $\varphi \in \mathcal{C}_0(\mathbb{R}^d \setminus \{0\})$ and prolonging it to zero with

$$\tilde{\varphi} = \begin{cases} \varphi(x), & x \neq 0 \\ 0 & x = 0, \end{cases}$$

we clearly have $\tilde{\varphi} \in \mathcal{C}_0(\mathbb{R}^d)$ and $\int w\tilde{\varphi}\,dx = \tilde{\varphi}(0) = 0$ which means that $\int w\varphi\,dx = 0$ for all $\varphi \in \mathcal{C}_c(\mathbb{R}^d \setminus \{0\})$. From this we conclude that $\tilde{A} = 0$, but on the other hand $\left\|\tilde{A}\right\|_{(L^\infty)'} = 1$, giving a contradiction (Table 8.1).

Reference

1. Brezis, H.E.: Functional Analysis, Sobolev Spaces and Partial Differential Equations, Universitext. Springer New York (2010)

Chapter 9
Elements of Spectral Theory

9.1 Invertible Operators

Let X be a Banach space over $\mathbb{K}$ (with $\mathbb{K} = \mathbb{C}$ or $\mathbb{K} = \mathbb{R}$) and consider the Banach space $\mathcal{L}(X)$ (recall the notation $\mathcal{L}(X) := \mathcal{L}(X, X)$) of linear bounded operators from X into itself. Given $T, S \in \mathcal{L}(X)$ also recall that we write $TS := T \circ S$ (composition operator) and we denote by $Id(\cdot)$ the identity operator $Id(x) = x$, for all $x \in X$. The space $\mathcal{L}(X)$ endowed with the operator sum, is a non-commutative algebra with unit.

Definition 9.1 (*Invertible operator*) Given $T \in \mathcal{L}(X)$ we say that T is invertible if there exists $S \in \mathcal{L}(X)$ such that $ST = TS = Id$, or equivalently if T is set-theoretically invertible: $\ker(T) = \{0\}$ and $\mathrm{Ran}(T) = X$. If T is invertible, the inverse is an operator in $\mathcal{L}(X)$ that we denote by T^{-1}.

Remark 9.2 Given T and S invertible elements in $\mathcal{L}(X)$, then TS is invertible too and $(TS)^{-1} = S^{-1}T^{-1}$. We use the conventional notation $T^0 = Id$, for all $T \in \mathcal{L}(X)$ and we observe that, for any T, S in $\mathcal{L}(X)$, we have

$$\|TS\| \leq \|T\| \, \|S\|, \tag{9.1}$$

where we recall Definition 3.4 of operator norm. Thus in particular, given two sequences $(T_n)_n$ and $(S_n)_n$ in $\mathcal{L}(X)$, such that $T_n \to T$ and $S_n \to S$ in $\mathcal{L}(X)$, then $T_n S_n \to TS$ in $\mathcal{L}(X)$.

From now on we will always assume $\mathbb{K} = \mathbb{C}$ if not differently specified.

Proposition 9.3 *Let $\mathcal{J} \subset \mathcal{L}(X)$ be the set of invertible operators. Then $\mathcal{J}$ is an open subset of $\mathcal{L}(X)$ containing the identity operator Id, and the map*

$$\mathcal{J} \ni T \to T^{-1} \tag{9.2}$$

is continuous. Moreover, if $T_0 \in \mathcal{J}$ and $T \in \mathcal{L}(X)$ is such that $\|T - T_0\| \leq \left\| T_0^{-1} \right\|^{-1}$, then

© The Author(s), under exclusive license to Springer Nature Switzerland AG 2026 107
S. Zagatti, *Functional Analysis*, SISSA Springer Series 7,
https://doi.org/10.1007/978-3-032-24475-8_9

$$T^{-1} := \sum_{n=0}^{+\infty} T_0^{-1}(Id - TT_0^{-1})^n = \sum_{n=0}^{+\infty}(Id - T_0^{-1}T)^n T_0^{-1}. \tag{9.3}$$

Proof Take $T_0 \in \mathcal{J}$. We first prove that $\mathcal{J}$ is open. To this aim, for any $T \in \mathcal{L}(X)$ we can write

$$T = T_0 - (T_0 - T) = T_0\big(Id - T_0^{-1}(T_0 - T)\big) = T_0(Id - A), \tag{9.4}$$

where $A := T_0^{-1}(T_0 - T)$. Remark that, using inequality (9.1), and supposing that $\|T - T_0\| \leq \|T_0^{-1}\|^{-1}$, we have

$$\|A\| \leq \|T_0^{-1}\| \, \|T_0^{-1}\|^{-1} = 1.$$

Thus, in order to conclude the proof, we now show that, if $A \in \mathcal{L}(X)$ and $\|A\| < 1$, then $(Id + A)$ is invertible. From identity (9.4) and Remark 9.2, this implies that $\mathcal{J}$ is open. To this aim, consider the formal series $\sum_{n=0}^{+\infty}(-1)^n A^n$ in $\mathcal{L}(X)$. First of all remark that

$$\left\|\sum_{n=m}^{m+p}(-1)^n A^n\right\| \leq \sum_{n=m}^{m+p} \|A\|^n \to 0, \quad \text{as } m \to \infty, \quad \forall p \in \mathbb{N}. \tag{9.5}$$

Thus, the series is convergent in $\mathcal{L}(X)$ and, by continuity of the composition map we have

$$(Id + A) \sum_{n=0}^{\infty}(-1)^n A^n = \sum_{n=0}^{\infty}(-1)^n A^n + \sum_{n=0}^{\infty}(-1)^n A^{n+1} = Id, \tag{9.6}$$

and analogously $\left(\sum_{n=0}^{\infty}(-1)^n A^n\right)(Id + A) = Id$. Thus $Id + A$ is invertible and

$$(Id + A)^{-1} = \sum_{n=0}^{\infty}(-1)^n A^n. \tag{9.7}$$

This proves that $\mathcal{J}$ is open and using (9.7) and (9.4), we obtain (9.3).

We conclude showing continuity of the inverse map (9.2). To do so, take any $T_0 \in \mathcal{J}$ and take T such that $\|T_0 - T\| \leq \|T_0^{-1}\|^{-1}$. We have

$$T^{-1} - T_0^{-1} \overset{(9.3)}{=} \sum_{n=0}^{\infty}(Id - T_0^{-1}T)^n T_0^{-1} - T_0^{-1} = \sum_{n=1}^{\infty}(Id - T_0^{-1}T)^n T_0^{-1}. \tag{9.8}$$

Thus

$$\left\| T^{-1} - T_0^{-1} \right\| \leq \left(\sum_{n=1}^{\infty} \left\| Id - T_0^{-1} T \right\|^n \right) \left\| T_0^{-1} \right\|$$

$$\leq \left\| T_0^{-1} \right\| \frac{\left\| Id - T_0^{-1} T \right\|}{1 - \left\| Id - T_0^{-1} T \right\|} \leq \frac{\left\| T_0^{-1} \right\|^2 \left\| T - T_0 \right\|}{1 - \left\| T_0^{-1} \right\| \left\| T - T_0 \right\|}, \tag{9.9}$$

where we have used the identity $Id - T_0^{-1} T = T_0^{-1}(T_0 - T)$. Thus, if $\| T - T_0 \| \to 0$, then $\left\| T^{-1} - T_0^{-1} \right\| \to 0$, obtaining continuity of the inverse map and concluding the proof. $\qquad\square$

9.2 Spectral Theory on Banach Spaces

Definition 9.4 (*Spectrum*) Let $\lambda \in \mathbb{C}$ and define $\lambda - T := \lambda Id - T$. If $\lambda - T$ is not invertible, we say that λ is a spectral value of T. We define the *spectrum* of T as

$$\sigma(T) := \{ \lambda \in \mathbb{C} : (\lambda - T) \text{ is not invertible} \}. \tag{9.10}$$

We denote by $\rho(T) := \mathbb{C} \setminus \sigma(T)$ the *resolvent set* of T.

Notation. We point out that the notation $\sigma(\cdot)$ used to denote the spectrum of an operator may overlap with the notation introduced in Chap. 7 for the topologies generated by families of functionals (see Definition 8.3). Nevertheless, since the notion of spectrum is used exclusively in Chap. 8, while induced topologies are discussed only in Chap. 7, this overlap should not cause any confusion.

Remark 9.5 Let $\lambda \in \mathbb{C}$ be such that $\ker(\lambda - T) \neq \{0\}$, then $\lambda \in \sigma(T)$ and we say that λ is an *eigenvalue* of T. We define the *point spectrum* of T as

$$\sigma_p(T) := \{ \lambda \in \sigma(T) : \lambda \text{ is an eigenvalue of } T \}. \tag{9.11}$$

To any $\lambda \in \sigma_p(T)$ we can associate the set $E_\lambda := \ker(\lambda - T)$, which is the *eigenspace* of λ, and any element $x \in E_\lambda$ (i.e., $Tx = \lambda x$) is an *eigenvector* associated with λ.

Proposition 9.6 *Let $T \in \mathcal{L}(X)$. Then the spectrum $\sigma(T)$ of T is a compact subset of $\mathbb{C}$ and*

$$\sigma(T) \subset \overline{B}^{\mathbb{C}}(0, r(T)), \tag{9.12}$$

where

$$r(T) := \lim_{n \to \infty} \left\| T^n \right\|^{\frac{1}{n}} = \inf_{n \in \mathbb{N}} \left\| T^n \right\|^{\frac{1}{n}}. \tag{9.13}$$

Proof Let $\lambda \in \rho(T)$. Since $\lambda - T \in \mathcal{J}$ and $\mathcal{J}$ is open in $\mathcal{L}(X)$ (see Proposition 9.3), there exists $\varepsilon > 0$ such that $\mu - T \in \mathcal{J}$ for all $\mu \in \mathbb{C}$ with $|\mu - \lambda| < \varepsilon$. Hence $\rho(T)$ is open. Thus, $\sigma(T)$ is a closed set and we are left to prove that $\lambda \in \sigma(T)$

implies $|\lambda| \leq r(T)$, with $r(T)$ given by (9.13). To this aim, take $\lambda \in \mathbb{C}$ such that $|\lambda| > r(T)$ and let $r > 0$ be such that $r(T) < r < |\lambda|$. This means that $\|T^n\|^{\frac{1}{n}} \leq r$ for n sufficiently big, and thus the series $\sum_{n=0}^{\infty} \lambda^{-n-1} T^n$ is convergent, since

$$\left\| \sum_{n=m}^{m+p} \lambda^{-n-1} T^n \right\| \leq \sum_{n=m}^{m+p} \frac{1}{|\lambda|} \left(\frac{r}{|\lambda|} \right)^n, \qquad \text{as } m \to \infty. \tag{9.14}$$

Performing calculations analogous to the ones in the proof of Proposition 9.3, one can check that the series converges to $(\lambda - T)^{-1}$. This shows that $(\lambda - T)^{-1} \in \mathcal{L}(X)$ and hence $\lambda \in \rho(T)$. We omit the details of these calculations.

We conclude the proof showing that the equality in (9.13) holds. To this aim, we first remark that $r(T) \leq \|T\|$. Next, take $\varepsilon > 0$ and $a := \inf_{n \in \mathbb{N}} \|T^n\|^{\frac{1}{n}}$ and let $n_0 \in \mathbb{N}$ be such that $\|T^{n_0}\|^{\frac{1}{n_0}} \leq a + \varepsilon$. Given any $n \in \mathbb{N}$ we write $n = p(n)n_0 + q(n)$, where $0 \leq q(n) < n_0$ and $p(n) \in \mathbb{N}$, so that

$$1 = \frac{p(n)}{n} n_0 + \frac{q(n)}{n} \quad \Longrightarrow \quad \frac{q(n)}{n} \to 0 \text{ and } \frac{p(n)}{n} \to \frac{1}{n_0}, \qquad \text{as } n \to +\infty. \tag{9.15}$$

We obtain

$$\left\| T^n \right\| = \left\| T^{p(n)n_0 + q(n)} \right\| \leq \left\| T^{n_0} \right\|^{p(n)} \|T\|^{q(n)}$$

and thus

$$\left\| T^n \right\|^{\frac{1}{n}} \leq \left\| T^{n_0} \right\|^{\frac{p(n)}{n}} \|T\|^{\frac{q(n)}{n}} \overset{(9.15)}{\to} \left\| T^{n_0} \right\|^{\frac{1}{n_0}} \leq a + \varepsilon.$$

From arbitrariness of ε we obtain

$$\limsup_{n \to \infty} \left\| T^n \right\|^{\frac{1}{n}} \leq a,$$

and since necessarily $\liminf \|T^n\|^{\frac{1}{n}} \geq a$, then the identity in (9.13) is verified, concluding the proof. $\square$

Definition 9.7 Let $T \in \mathcal{L}(X)$ and $\lambda \in \rho(T)$. The operator $R(\lambda, T) := (\lambda - T)^{-1}$ is called *resolvent operator* (or simply resolvent) of T at λ.

Lemma 9.8 (Resolvent Formula)
   Let $T \in \mathcal{L}(X)$. Then for every $\lambda, \mu \in \rho(T)$ we have

$$R(\lambda, T) - R(\mu, T) = (\mu - \lambda)R(\lambda, T)R(\mu, T). \tag{9.16}$$

Moreover, the map $\rho(T) \ni \lambda \to R(\lambda, T)$ is differentiable and we have

$$\frac{\mathrm{d}}{\mathrm{d}\lambda} R(\lambda, T) = -R(\lambda, T)^2. \tag{9.17}$$

Proof In order to prove (9.16) we first observe that $R(\lambda, T)$ commutes with $\lambda - T$ and hence with any operator of the form $\mu - T$, with $\mu \in \mathbb{C}$, since $\mu - T = (\lambda - T) + (\mu - \lambda)Id$. In particular we have

$$R(\lambda, T) = R(\lambda, T)(\mu - T)R(\mu, T) = (\mu - T)R(\lambda, T)R(\mu, T) \qquad (9.18)$$

and we can also write

$$R(\mu, T) = (\lambda - T)R(\lambda, T)R(\mu, T). \qquad (9.19)$$

Subtracting (9.19) from (9.18) we obtain (9.16).

Finally, we use Proposition 9.3 and in particular continuity of the map $\rho(T) \ni \lambda \to R(\lambda, T)$ and we have

$$\frac{R(\lambda, T) - R(\mu, T)}{\lambda - \mu} \overset{(9.16)}{=} -R(\lambda, T)R(\mu, T) \to -R(\lambda, T)^2, \quad \text{as } \mu \to \lambda,$$
$$(9.20)$$

concluding the proof. $\qquad\qquad\qquad\qquad\qquad\qquad\qquad\qquad\qquad\qquad\qquad\quad\square$

Proposition 9.9 (Spectral radius) *Let* $T \in \mathcal{L}(X)$. *Then* $\sigma(T) \neq \emptyset$ *and* $\max\{|\lambda| : \lambda \in \sigma(T)\} = r(T)$ *where* $r(T)$ *is defined in* (9.13) *and is called the spectral radius of* T.

Proof We proceed in several steps. First let $\lambda \in \mathbb{C}$ and suppose that $|\lambda| > r(T)$. Then, using the definition of $R(\lambda, T)$ in 9.7 and applying Proposition 9.3, we get $R(\lambda, T) = \sum_{n=0}^{\infty} \lambda^{-n-1} T^n$. Moreover, rewriting λ as $\lambda = te^{i\theta}$, with $t > 0$ and $\theta \in [0, 2\pi]$, we obtain

$$R(\lambda, T) = R(te^{i\theta}, T) = \sum_{n=0}^{\infty} t^{-(n+1)} e^{-i(n+1)\theta} T^n.$$

Thus, for any $p \in \mathbb{N}_0 = \mathbb{N} \cup \{0\}$, we have

$$\lambda^{p+1} R(\lambda, T) = t^{p+1} e^{i(p+1)\theta} R(te^{i\theta}, T) = \sum_{n=0}^{\infty} t^{p-n} e^{i(p-n)\theta} T^n. \qquad (9.21)$$

Integrating (9.21) with respect to θ between 0 and 2π we obtain

$$T^p = \frac{1}{2\pi} \int_0^{2\pi} (te^{i\theta})^{p+1} R(te^{i\theta}, T) \, d\theta. \qquad (9.22)$$

Having introduced these notations, we now prove that $\sigma(T) \neq \emptyset$. We suppose by contradiction that $\sigma(T) = \emptyset$ (or equivalently that $\rho(T) = \mathbb{C}$) and we consider the function $J_0 : [0, +\infty[\to \mathcal{L}(X)$ defined by

$$J_0(t) := \frac{1}{2\pi} \int_0^{2\pi} (te^{i\theta}) R(te^{i\theta}, T) d\theta. \qquad (9.23)$$

Since $\rho(T) = \mathbb{C}$, we see that $J_0(\cdot)$ is a continuous function over the whole $[0, +\infty[$. Moreover, from (9.22) we also have $J_0(t) = Id$ for any $t > 0$. By continuity of J_0, we would then have $J_0(0) = Id$, but from the integral formula (9.23) we see that $J_0(0) = 0$. This is a contradiction, thus $\sigma(T) \neq \emptyset$.

Next, we show that $\max\{|\lambda| : \lambda \in \sigma(T)\} = r(T)$. Reasoning again by contradiction, suppose that $\alpha := \max\{|\lambda|, \lambda \in \sigma(T)\} < r(T)$ and consider the functions

$$J_n(t) := \frac{1}{2\pi} \int_0^{2\pi} (te^{i\theta})^{n+1} R(te^{i\theta}, T)\mathrm{d}\theta, \quad n \in \mathbb{N}, t > \alpha. \tag{9.24}$$

Since $J_n(t) = T^n$ if $t > r(T)$, we now show that $\frac{\mathrm{d}}{\mathrm{d}t} J_n(t) = 0$ for any $t \geq \alpha$, obtaining that $J_n(t) = T^n$ for any $T \in [\alpha, +\infty[$. This produces a contradiction since, from the definition of $J_n(t)$ in (9.24) we also have

$$\|T^n\| \leq t^{n+1} \max\{\|R(te^{i\theta}, T)\|, \theta \in [0, 2\pi]\}, \quad \forall n \in \mathbb{N}, t > \alpha,$$

and thus $r(T) = \lim_{n \to \infty} \|T^n\|^{\frac{1}{n}} \leq t$ for any $t > \alpha$, contradicting the assumption $\alpha < r(T)$.

We are left to prove that $\frac{\mathrm{d}}{\mathrm{d}t} J_n(t) = 0$ in for any $t \geq \alpha$:

$$\begin{aligned}
\frac{\mathrm{d}}{\mathrm{d}t} J_n(t) &= \frac{1}{2\pi} \int_0^{2\pi} \partial_t \left[(te^{i\theta})^{n+1} R(te^{i\theta}, T) \right] \mathrm{d}\theta \\
&= \frac{1}{2\pi i t} \int_0^{2\pi} \partial_\theta \left[(te^{i\theta})^{n+1} R(te^{i\theta}, T) \right] \mathrm{d}\theta = 0,
\end{aligned} \tag{9.25}$$

where we have used the identities

$$\partial_t \left[(te^{i\theta})^{n+1} R(te^{i\theta}, T) \right] = \frac{\mathrm{d}}{\mathrm{d}z} \left[z^{n+1} R(z, T) \right]\big|_{z=te^{i\theta}} (n+1)e^{i\theta(n+1)} \tag{9.26}$$

and

$$\partial_\theta \left[(te^{i\theta})^{n+1} R(te^{i\theta}, T) \right] = \frac{\mathrm{d}}{\mathrm{d}z} \left[z^{n+1} R(z, T) \right]\big|_{z=te^{i\theta}} t^{n+1} i(n+1)e^{i\theta(n+1)}. \tag{9.27}$$

$\square$

Let $P(y) := \sum_{k=0}^m a_k y^k$, with $a_k \in \mathbb{C}$, a polynomial with complex coefficients. For any $T \in \mathcal{L}(X)$ the operator $P(T) := \sum_{k=0}^m a_k T^k$ is well defined and belongs to $\mathcal{L}(X)$. We have the following result.

Lemma 9.10 (Spectral mapping for polynomials) *Let* $T \in \mathcal{L}(X)$, *with* X *Banach space over* $\mathbb{C}$. *Then*

$$P(\sigma(T)) = \{p(\lambda) : \lambda \in \sigma(T)\} = \sigma(P(T)). \tag{9.28}$$

Remark 9.11 If the Banach space X is real, only the inclusion $P(\sigma(T)) \subset \sigma(P(T))$ holds in general.

Proof We first take $\lambda \in \sigma(T)$, consider $P(\lambda) \in P(\sigma(T))$ and prove that $P(\lambda) \in \sigma(P(T))$. To this aim, notice that the polynomial $y \to P(y) - P(\lambda)$ has a zero in $y = \lambda$, thus we can write

$$P(y) - P(\lambda) = (y - \lambda)Q_\lambda(y),$$

for some polynomial $Q_\lambda(\cdot)$. Thus, using that a polynomial in T commutes with T, we have:

$$P(T) - P(\lambda) = (T - \lambda)Q_\lambda(T) = Q_\lambda(T)(T - \lambda).$$

Suppose now by contradiction that $P(\lambda) \notin \sigma(P(T))$ and let $S := (P(T) - P(\lambda))^{-1}$. Then $Id = (T - \lambda)Q_\lambda(T)S = SQ_\lambda(T)(T - \lambda)$, and thus $T - \lambda$ is invertible. This contradicts the fact that $\lambda \in \sigma(T)$, concluding the proof.

We now prove that $\sigma(P(T)) \subset P(\sigma(T))$. We suppose that P has degree m greater or equal to one, otherwise the thesis is trivially true. Take $\mu \in \sigma(P(T))$. Using the fundamental theorem of algebra we can decompose $P(y) - \mu$ as

$$P(y) - \mu = c(y - \lambda_1)\ldots(y - \lambda_m), \quad 0 \neq c \in \mathbb{C}, \ \lambda_1, \ldots, \lambda_m \in \mathbb{C}.$$

Thus,

$$P(T) - \mu = c(T - \lambda_1)\ldots(T - \lambda_m).$$

If $\lambda_j \in \rho(T)$ for all $j = 1, \ldots, m$, then $P(T) - \mu$ would be invertible. This would imply $\mu \notin \sigma(P(T))$, contradicting our assumption. Thus, there exists some $j \in \{1, \ldots, m\}$ such that $\lambda_j \in \sigma(T)$, and being $\mu = P(\lambda_j)$, then $\mu \in P(\sigma(T))$ concluding the proof. $\qquad\square$

9.3 Spectral Theory on Hilbert Spaces

In this section, we study the spectrum of operators on $\mathcal{L}(\mathcal{H})$, where $\mathcal{H}$ is a Hilbert space that we will assume to be complex. Since we will extensively study the spectrum of the adjoint operator T^* and properties of the spectrum of selfadjoint operators (see Definition 5.27), we first recall the following result proved in Section 5.5.

Proposition 9.12 *Let $\mathcal{H}$ be a Hilbert space, and let $T \in \mathcal{L}(\mathcal{H})$. Then there exists a unique $T^* \in \mathcal{L}(\mathcal{H})$ such that $(Tx, y) = (x, T^*y)$ for all $x, y \in H$. Moreover, we have $\|T\| = \|T^*\|$. We call T^* the adjoint operator of T.*

Moreover, using the definition of the adjoint operator and the properties of Hilbert spaces one can immediately prove the following properties.

Lemma 9.13 (Properties of the adjoint operator) *Let $S, T \in \mathcal{L}(\mathcal{H})$ and $\alpha \in \mathbb{C}$. Then*

1. $(T + S)^* = T^* + S^*$;
2. $(\alpha T)^* = \overline{\alpha} T^*$;
3. $(ST)^* = T^* S^*$;
4. $(T^*)^* = T$;
5. $Id^* = Id$.

Lemma 9.14 *Let $T \in \mathcal{L}(\mathcal{H})$. Then $\|TT^*\| = \|T^*T\| = \|T\|^2$.*

Proof From (9.1) we immediately have that $\|TT^*\| \leq \|T\|^2$ and $\|T^*T\| \leq \|T\|^2$. On the other hand, we also have

$$\|Tx\|^2 = (Tx, Tx) = (x, T^*Tx) \leq \|x\|^2 \|T^*T\|,$$

which implies $\|T\|^2 \leq \|T^*T\|$. Analogously

$$\left\|T^*x\right\|^2 = (T^*x, T^*x) = (TT^*x, x) \leq \|x\|^2 \left\|TT^*\right\|,$$

which implies $\|T\|^2 \leq \|TT^*\|$ concluding the proof. $\qquad\qquad\square$

Lemma 9.15 *Let $T \in \mathcal{L}(\mathcal{H})$. Then*

1. $\ker(T) = (Ran\ T^*)^{\perp}$;
2. $\overline{Ran\ T} = (\ker(T^*))^{\perp}$;
3. $T \in \mathcal{J} \iff T^* \in \mathcal{J}$ and $(T^*)^{-1} = (T^{-1})^*$,

where we recall the notation $\mathcal{J}$ for the set of invertible operators in $\mathcal{L}(\mathcal{H})$ and the Definition 5.13 of the orthogonal complement.

Proof We prove separately the three points.

1. Let $x \in \ker(T)$, then $(x, T^*y) = 0$ for every $y \in \mathcal{H}$. But since

$$(x, T^*y) = 0 \text{ for every } y \in \mathcal{H} \iff x \in (Ran\ T^*)^{\perp},$$

   we have proved the first point.
2. To prove the second point it is enough to observe that $\overline{Ran\ T} = ((Ran\ T^*)^{\perp})^{\perp} = \ker(T)^{\perp}$, which proves the claim recalling that $(T^*)^* = T$.
3. Writing $TT^{-1} = Id = T^{-1}T$, we get

$$(TT^{-1})^* = (T^{-1})^*T^* = Id^* = Id \quad \text{and} \quad (T^{-1}T)^* = T^*(T^{-1})^* = Id^* = Id.$$

$$\qquad\qquad\square$$

Corollary 9.16 *Let $T \in \mathcal{L}(\mathcal{H})$. Then $\sigma(T^*) = \{\overline{\lambda} : \lambda \in \sigma(T)\}$ and $\rho(T^*) = \{\overline{\lambda} : \lambda \in \rho(T)\}$. Moreover, for every $\lambda \in \rho(T)$ we have $R(\lambda, T)^* = R(\overline{\lambda}, T^*)$.*

Lemma 9.17 *Let $T \in \mathcal{L}(\mathcal{H})$ be selfadjoint. Then $r(T) = \|T\|$.*

In particular, for any $S \in \mathcal{L}(\mathcal{H})$ (not necessarily selfadjoint) $\|S\| = r(S^*S)^{\frac{1}{2}} = r(SS^*)^{\frac{1}{2}}$.

Proof Recall from Lemma 9.14 that $\|T^2\| = \|T^*T\| = \|T\|^2$. Thus in particular, for every $n \in \mathbb{N}$ we have $\|T^{2^n}\| = \|T\|^{2^n}$ and

$$r(T) \overset{(9.13)}{=} \lim_{n \to \infty} \|T^{2^n}\|^{\frac{1}{2^n}} = \|T\| \, ,$$

concluding the first part of the proof.

Take now any $S \in \mathcal{L}(\mathcal{H})$ and remark that S^*S and SS^* are both selfadjoint operators. The second claim thus follows immediately from the first, with $T = S^*S$ and $T = SS^*$. $\qquad\square$

Proposition 9.18 (Spectrum of selfadjoint operators) *Let $T \in \mathcal{L}(\mathcal{H})$ be a selfadjoint operator. Then*

1. $\sigma_p(T) \subset \mathbb{R}$ *(recall (9.11))*;
2. *for any $\lambda \in \mathbb{C}$, $\overline{Ran(\lambda - T)} = \ker(\bar{\lambda} - T)^{\perp}$*;
3. *given any $\lambda, \mu \in \sigma_p(T)$, with $\lambda \neq \mu$, and any $x \in \ker(\lambda - T)$ and $y \in \ker(\mu - T)$, then $x \perp y$.*

Proof We prove separately each point.

1. Let $\lambda \in \sigma_p(T)$. Then there exists $x \in \mathcal{H}$, $x \neq 0$, such that $Tx = \lambda x$. We have from direct calculation that (Tx, x) is real valued, since

$$(Tx, x) = (x, T^*x) = (x, Tx) = \overline{(Tx, x)},$$

    but also

$$(Tx, x) = \lambda(x, x) = \lambda \|x\|^2 \, .$$

    Thus $\lambda \in \mathbb{R}$, concluding the proof of this first point.
2. This claim directly follows from Lemma 9.15.
3. Take λ, μ distinct eigenvalues and x and y corresponding eigenvectors. Then:

$$(Tx, y) = (x, Ty) \implies \lambda(x, y) = \mu(x, y) \implies (x, y) = 0,$$

    proving orthogonality of x and y.

$\qquad\square$

Proposition 9.19 *Let $T \in \mathcal{L}(\mathcal{H})$ be selfadjoint and let*

$$m := \inf_{\|x\|=1} (Tx, x); \quad and \quad M := \sup_{\|x\|=1} (Tx, x). \tag{9.29}$$

Then $\sigma(T) \subset [m, M]$ and $m, M \in \sigma(T)$.

To prove the proposition we need the following preliminary definition.

Definition 9.20 Let $S \in \mathcal{L}(\mathcal{H})$, we say that S is a positive operator if, for any $x \in \mathcal{H}$, $(Sx, x) \geq 0$.

We are now in position to prove Proposition 9.19.

Proof Let $\xi \in \mathcal{H}$, $\|\xi\| = 1$ and for every $\lambda \in \mathbb{C}$ define the quantity

$$d(\lambda) := \min\{|\lambda - t| : t \in [m, M]\}.$$

Remark that $d(\lambda) = d(\overline{\lambda})$. Then we have:

$$|(\lambda \xi - T\xi, \xi)| = |(\lambda \xi, \xi) - (T\xi, \xi)| = |\lambda - (T\xi, \xi)| \geq d(\lambda).$$

Analogously, for every $x \in \mathcal{H}$ we have

$$\|(\lambda - T)x\| \geq d(\lambda) \|x\|. \tag{9.30}$$

Take now any $\lambda \notin [m, M]$, i.e., $d(\lambda) > 0$. Then, from (9.30), $(\lambda - T)x = 0$ if and only if $x = 0$, or equivalently $\ker(\lambda - T) = \{0\}$. We now show that $\mathrm{Ran}(\lambda - T)$ is closed, proving that $\lambda \in \rho(T)$. Let $y_n = (\lambda - T)x_n \to y$. Using (9.30) we obtain $\|y_n - y_m\| \geq d(\lambda) \|x_n - x_m\|$ and thus $x_n \to x$ for some $x \in \mathcal{H}$ and $y = (\lambda - T)x \in \mathrm{Ran}(\lambda - T)$.

On the other hand, $d(\lambda) = d(\overline{\lambda})$, and thus $\ker(\overline{\lambda} - T) = \{0\}$ and

$$\overline{\mathrm{Ran}(\lambda - T)} = \mathrm{Ran}(\lambda - T) = \ker(\overline{\lambda} - T) = \{0\}^{\perp} = \mathcal{H}.$$

Thus, $\lambda - T$ is set-theoretically invertible and from (9.30) we have $\left\|(\lambda - T)^{-1}\right\| \leq d(\lambda)^{-1}$. Thus, $\lambda \in \rho(T)$, proving that $\sigma(T) \subseteq [m, M]$. We are left to prove that $m, M \in \sigma(T)$. The two proofs are identical, thus, we only show that $m \in \sigma(T)$.

To this aim, define $S := T - mId$ and remark that S is a selfadjoint operator and it is also positive in the sense of Definition 9.20. Indeed,

$$(Sx, x) = (Tx - mx, x) = (Tx, x) - m(x, x) = \|x\|^2 \left[\left(T\frac{x}{\|x\|}, \frac{x}{\|x\|}\right) - m\right] \overset{(9.29)}{\geq} 0. \tag{9.31}$$

Thus, thanks to this positivity property one can prove that the map

$$\mathcal{H} \times \mathcal{H} \ni \{x, y\} \to (Sx, y)$$

is a semi-scalar product, and, write the associated Cauchy-Schwarz inequality:

$$|(Sx, y)| \leq (Sx, x)^{\frac{1}{2}} (Sy, y)^{\frac{1}{2}}, \quad \forall x, y \in \mathcal{H}. \tag{9.32}$$

We conclude using the definition of S and the properties of m: we prove that there exists a sequence $(x_n)_n$ in $\mathcal{H}$, $\|x_n\| = 1$ and such that $Sx_n \to 0$. If we suppose by contradiction that $m \notin \sigma(T)$, then S is invertible and $x_n = S^{-1}Sx_n \to 0$, contradicting the hypothesis that $\|x_n\| = 1$, and thus concluding the proof.

To find the sequence $(x_n)_n$, we use the definition of m: there exists $(x_n)_n$, $\|x_n\| = 1$ and such that $(Sx_n, x_n) \to 0$, as $n \to +\infty$. Moreover, we have

$$\|Sx_n\|^2 = (Sx_n, Sx_n) \leq (Sx_n, x_n)^{\frac{1}{2}}(S^2 x_n, Sx_n)^{\frac{1}{2}}$$

$$\leq (Sx_n, x_n)^{\frac{1}{2}}\left\|S^2 x_n\right\|^{\frac{1}{2}}\|Sx_n\|^{\frac{1}{2}} \tag{9.33}$$

$$\leq (Sx_n, x_n)^{\frac{1}{2}}\|S\|^{\frac{1}{2}}\|Sx_n\|,$$

where in the first inequality we have applied the Cauchy-Schwarz inequality (9.32) and positivity of the operator S.

Thus $\|Sx_n\| \leq \|S\|^{\frac{1}{2}}(Sx_n, x_n)^{\frac{1}{2}}$ and in particular $Sx_n \to 0$, concluding the proof. $\qquad\square$

Corollary 9.21 *Let $T \in \mathcal{L}(\mathcal{H})$ be a selfadjoint operator. Then $\|T\| = \max\{|m|, |M|\}$.*

Proof The Corollary immediately follows from the fact that $\|T\| = r(T) = \max\{|\lambda| : \lambda \in \sigma(T)\}$, and applying Proposition 9.19. $\qquad\square$

Corollary 9.22 *Let $T \in \mathcal{L}(\mathcal{H})$ be selfadjoint. Then T is positive if and only if $\sigma(T) \subseteq [0, +\infty[$. In this case $\|T\| \in \sigma(T)$.*

Let now $P(y) = \sum_{k=0}^{m} a_k y^k$ be a polynomial of degree m with complex coefficients $a_k \in \mathbb{C}$, $k = 0, \ldots, m$, and let $\overline{P}(y) = \sum_{k=0}^{m} \overline{a}_k y^k$. We have the following result.

Theorem 9.23 *Let $T \in \mathcal{L}(\mathcal{H})$ be a selfadjoint operator and let P be the polynomial defined above. Then $P(T)^* = \overline{P}(T)$ and $\|P(T)\| = \max_{\lambda \in \sigma(T)} |P(\lambda)|$.*

Proof The first identity immediately follows from the fact that, for every $\alpha \in \mathbb{C}$ we have $(\alpha T)^* = \overline{\alpha}T^* = \overline{\alpha}T$ and that $(T^n)^* = (T^*)^n = T^n$, for every $n \in \mathbb{N}$. In order to prove the second identity, remark that, from Lemma 9.14, we have

$$\|P(T)\|^2 = \left\|P(T)^*P(T)\right\| = \left\|\overline{P}(T)P(T)\right\| = \left\||P|^2(T)\right\|,$$

where $|P|^2(y) = P(y)\overline{P}(y) = |P(y)|^2$, for all $y \in \mathbb{R}$. One can check that $|P|^2(T)$ is selfadjoint and positive. Moreover, from Lemma 9.10 we have $\sigma(|P|^2(T)) = |P|^2(\sigma(T))$. In particular we have

$$\|P(T)\|^2 = \max\{|\mu| : \mu \in \sigma(|P|^2(T))\} = \max\{|P|^2(\lambda) : \lambda \in \sigma(T)\} = \left(\max_{\lambda \in \sigma(T)} |P(\lambda)|\right)^2, \tag{9.34}$$

concluding the proof. $\qquad\square$

Theorem 9.24 (Functional Calculus for selfadjoint operators) *Let $\mathcal{H}$ be a Hilbert space over $\mathbb{C}$ and let $T \in \mathcal{L}(\mathcal{H})$ be a selfadjoint operator. Then there exists a linear isometry $\phi : C(\sigma(T)) \to \mathcal{L}(\mathcal{H})$, such that, defining $C(\sigma(T)) \ni f \to \phi(f) = f(T)$, the following identities hold:*

1. $(fg)(T) = f(T)g(T), \quad \forall f, g \in C(\sigma(T));$
2. $f(T)^* = \overline{f}(T), \quad \forall f \in C(\sigma(T));$
3. $\sigma(f(T)) = f(\sigma(T)), \quad \forall f \in C(\sigma(T)).$

Proof We prove the result in several steps. Define $\Pi := \{P|_{\sigma(T)}, P \text{ polynomial }\}$. We know from Theorem 9.23 that the map $\Pi \ni P \to P(T) \in \mathcal{L}(\mathcal{H})$ is a linear isometry, since $\|P(T)\| = \|P\|_{C(\sigma(T))}$. From the Stone-Weierstrass Theorem 4.12, Π is dense in $C(\sigma(T))$ and thus we can apply the extension Theorem 2.7, obtaining that there exists $\phi : C(\sigma(T)) \to \mathcal{L}(\mathcal{H})$, linear isometry, and consequently define $f(T) := \phi(f)$, for all $f \in C(\sigma(T))$. Given $f \in C(\sigma(T))$ and $(P_n)_n$ sequence in Π such that $P_n \to f$ uniformly, we have $f(T) = \lim_{n \to \infty} P_n(T)$ (where the limit is taken in $\mathcal{L}(\mathcal{H})$). Thus, we obtain properties 1 and 2 on the whole $C(\sigma(T))$ since they hold for all $P_n \in \Pi$ and are preserved when taking the limit by continuity of the map $(S, T) \to ST$ and $T \to T^*$.

We now prove the spectral mapping property in the third item. We first show that $\sigma(f(T)) \subseteq f(\sigma(T))$. To this aim, let $\lambda \notin f(\sigma(T))$, so that $\frac{1}{\lambda - f} \in C(\sigma(T))$, since the denominator never vanishes in $\sigma(T)$. Moreover, since $\frac{1}{\lambda - f}(\lambda - f) = 1$, $(\lambda - f)\frac{1}{\lambda - f} = 1$, using the first item we have

$$\frac{1}{\lambda - f}(T)(\lambda - f)(T) = (\lambda - f)(T)\frac{1}{\lambda - f}(T) = Id.$$

Thus $\lambda - f(T)$ is invertible and $\lambda \in \rho(f(T))$, concluding the proof of the first inclusion.

In order to prove the other inclusion, take $f \in C(\sigma(T))$, $f \geq 0$ and $f(T)$ invertible. From previous step, $\sigma(f(T)) \subseteq f(\sigma(T)) \overset{(9.22)}{\subseteq} [0, +\infty[$ and thus $]-\infty, 0] \subseteq \rho(f(T))$. In particular, $-\frac{1}{n} \in \rho(f(T))$ for any $n \in \mathbb{N}$ and $0 \in \rho(f(T))$. Using the first property, we obtain $R(-\frac{1}{n}, f(T)) = \frac{1}{(-\frac{1}{n} - f)}(T)$ and thus

$$\left\| R\left(-\frac{1}{n}, f(T)\right) \right\| = \left\| \frac{1}{-\frac{1}{n} - f} \right\|_{C(\sigma(T))}. \tag{9.35}$$

By continuity of the map $\rho(T) \ni \lambda \to R(\lambda, T)$ we have $R(-\frac{1}{n}, f(T)) \to R(0, f(T))$ in $\mathcal{L}(\mathcal{H})$ and $R(0, f(T)) = f^{-1}(T) = \frac{1}{f}(T)$. We conclude that the sequence $(\|R(-\frac{1}{n}, f(T))\|)_n$ is bounded in $\mathbb{R}$ and from (9.35) this implies that f cannot vanish in $\sigma(T)$, thus $0 \notin f(T)$. This proves the third item for any non-negative continuous function.

We conclude extending to any function $f \in C(\sigma(T))$: take any $\lambda \in \rho(f(T))$. If we show that $\lambda \notin f(\sigma(T))$ we can deduce that $f(\sigma(T)) \subseteq \sigma(f(T))$. Since $(\lambda - f(T))$

is invertible, also $(\lambda - f(T))^* = (\overline{\lambda} - \overline{f}(T))$ is, and thus $|\lambda - f|^2(T)$ is invertible. Since $|\lambda - f|^2 \geq 0$ and $|\lambda - f|^2(T)$ is invertible, we can apply the previous step to $|\lambda - f|^2$, concluding that $|\lambda - f|^2$ does not vanish on $\sigma(T)$, and hence $\lambda \notin f(\sigma(T))$. $\qquad\square$

9.4 Spectral Decomposition of Compact Selfadjoint Operators

In this section we prove some properties of the spectrum of compact operators, referring to Section 3.6 for the definition of compact operators and some general results that we will use in the following.

Lemma 9.25 *Let $\mathcal{H}$ be a Hilbert space and let $T \in \mathcal{L}(\mathcal{H})$ be a compact operator. Then*

1. *$\ker(Id - T)$ is finite dimensional;*
2. *$Ran(Id - T)$ is closed;*
3. *$Id - T$ is invertible if and only if it is an injective map.*

Proof We prove separately each item.

1. Let $F := \ker(Id - T)$. F is a subspace of $\mathcal{H}$ such that $x \in F$ if and only if $x = Tx$. Then in particular $\overline{B}_1^F = T(\overline{B}_1^F) \subset \overline{T(\overline{B}_1^{\mathcal{H}})} \cap F$. Since T is compact, then also $\overline{T(\overline{B}_1^{\mathcal{H}})}$ is, and thus $\overline{B}_1^F$ is compact. From Theorem 2.27 we conclude that F is also finite dimensional.

2. Let $y_n := (Id - T)x_n \to y$, as $n \to \infty$, with $(x_n)_n$ sequence in $\mathcal{H}$. We consider first the case in which $(x_n)_n$ is a bounded sequence. Using compactness of T, up to subsequence we can assume that $Tx_n \to z$. Thus $x_n - Tx_n \to y - z$ and $x_n \to y + z$. Hence, we conclude that $T(y + z) = z$ and $(y + z) - T(y + z) = y$, proving that $y \in Ran(Id - T)$.

   Consider now the case in which $(x_n)_n$ is unbounded, and let $z_n \in F = \ker(Id - T)$ be such that $\|x_n - z_n\| = \mathrm{dist}(x_n, F) =: d_n$. Suppose first that $(d_n)_n$ is bounded, then up to subsequence we have $T(x_n - z_n) \to z$ and thus, being $z_n = Tz_n$, also $(x_n - z_n) - T(x_n - z_n) \to y$ and $x_n - z_n \to y + z$. Thus, $T(y + z) = z$ and $(y + z) - T(y + z) = y$ and we can conclude as in the previous case.

   Finally, suppose that $(d_n)_n$ is unbounded. Reasoning as in previous case we arrive to $T\left(\frac{x_n - z_n}{d_n}\right) \to u$ and since $z_n = Tz_n$ we can infer $\frac{x_n - z_n}{d_n} \to u$ and thus $Tu = u$, or equivalently $u \in F$. On the other hand, taking n large enough, we have $\left\|\frac{x_n - z_n}{d_n} - u\right\| \leq \frac{1}{2}$, which is equivalent to $\|x_n - (z_n + d_n u)\| \leq \frac{d_n}{2}$, but this is impossible since $z_n + d_n u \in F$. Thus, $(d_n)_n$ is bounded and this concludes the proof of the second point.

3. To prove this equivalence, it is enough to show that if $Id - T$ is injective, then $Id - T$ is invertible. First, we show that $\mathrm{Ran}(Id - T) = \mathcal{H}$. To this aim we pose $E_0 := \mathcal{H}$ and $E_1 := (Id - T)E_0$. For any $n \in \mathbb{N} \cup \{0\}$, iterating, we define $E_{n+1} := (Id - T)E_n$. Suppose by contradiction that $E_1 \subsetneq E_0$. We show, by induction, that E_n is closed and $E_{n+1} \subsetneq E_n$, for any $n \in \mathbb{N} \cup \{0\}$. Suppose that $E_n \subsetneq E_{n-1}$, then $(Id - T)(E_n) = E_n - T(E_n) = E_{n+1} \subsetneq E_n$. Thus, we have $T(E_n) \subseteq E_n$. Since E_n is closed, $T_n := T|_{E_n} \in \mathcal{L}(E_n)$. Moreover, we also have $T_n\left(\overline{B}_1^{E_n}\right) \subseteq \overline{T(\overline{B}_1^{\mathcal{H}})} \cap E_n$, which is compact. This implies that T_n is compact and thus, from the second point of the theorem, we have

$$E_n = (Id - T)E_n = \mathrm{Ran}(Id - T_n)$$

which is closed. Finally, we have

$$E_{n+2} = (Id - T)E_{n+1} \subsetneq (Id - T)E_n \subseteq E_{n+1},$$

proving the desired property.

For $n \in \mathbb{N} \cup \{0\}$ let now $u_n \in E_n$, $\|u_n\| = 1$, such that $\mathrm{dist}(u_n, E_{n+1}) \geq \frac{1}{2}$. Remark that u_n exists since E_{n+1} is a closed subspace, strictly contained in E_n. Given $m, n \in \mathbb{N}$, $m > n$, we write

$$Tu_n - Tu_m = u_m - v_{n,m}, \quad \text{where} \quad v_{n,m} := (Id - T)u_n + Tu_m.$$

Remark that $Tv_m \in T(E_m) \subset E_m \subsetneq E_{n+1}$, since $m > n$, and $(Id - T)u_n \in E_{n+1}$, since $u_n \in E_n$. Thus

$$\|Tu_n - Tu_m\| = \left\|u_n - v_{n,m}\right\| \geq \frac{1}{2}.$$

Finally, since $\|u_n\| = 1$ for any $n \in \mathbb{N}$, this last inequality contradicts compactness of T. This concludes the proof proving that $Id - T$ is surjective.

Thus we can now denote by $(Id - T)^{-1}$ the set-theoretic inverse of $(Id - T)$, and we are left to show that it is continuous to conclude the proof. Suppose by contradiction that there exists $(y_n)_{n\in\mathbb{N}}$ in $\mathcal{H}$ such that $\|(Id - T)y_n\| \, \|y_n\|^{-1} \to \infty$, as $n \to \infty$. Denote by $x_n := (Id - T)^{-1}y_n$ and thus $y_n = (Id - T)x_n$, so that $\|x_n\| \, \|(Id - T)x_n\|^{-1} \to \infty$ as $n \to \infty$. Thus introducing $\xi_n := \frac{x_n}{\|x_n\|}$, we obtain

$$\frac{1}{\|(Id - T)\xi_n\|} \to \infty \quad \Longrightarrow \quad (Id - T)\xi_n \to 0. \tag{9.36}$$

Since T is compact, up to subsequence we have $T\xi_n \to v$ and $\xi_n \to v$ for some $v \in \mathcal{H}$ with $\|v\| = 1$ and thus $(Id - T)v = 0$, contradicting injectivity of v.

$\square$

Proposition 9.26 *Let $\mathcal{H}$ be a Hilbert space and let $T \in \mathcal{L}(\mathcal{H})$ be a compact operator. Then*

1. *for every $\lambda \in \sigma(T) \setminus \{0\}$, $\lambda \in \sigma_p(T)$ and the eigenspace E_λ associated with λ is finite dimensional;*
2. *the set $\sigma_p(T)$ is countable and, when it is not finite, it consists of sequence $(\lambda_n)_n$, decreasing in modulus, and which converges to zero.*

Proof We prove separately the two items.

1. Let $\lambda \in \mathbb{C} \setminus \{0\}$ and write $\lambda - T = \lambda\left(Id - \lambda^{-1}T\right)$. Using now the third point of Lemma 9.25, we see that $\lambda - T$ is invertible if and only if $\left(Id - \lambda^{-1}T\right)$ is injective. Thus $\lambda \in \sigma(T)$ if and only if $\lambda \in \sigma_p(T)$. Moreover, using the first point of Lemma 9.25, we also conclude that $E_\lambda = \ker(\lambda - T) = \ker\left(I - \lambda^{-1}T\right)$ is finite dimensional.

2. To prove this second item we suppose by contradiction that there exists $\varepsilon > 0$ such that the set $U_\varepsilon := \{\lambda \in \sigma(T) : |\lambda| \geq \varepsilon\}$ is infinite. Showing that this condition cannot hold completes the proof of the second point of the proposition, since it implies that

$$\sigma_p(T) = \bigcup_{n=1}^{\infty} \{\lambda \in \sigma(T) : |\lambda| \geq \frac{1}{n}\}$$

is countable, since every set of the form $\{\lambda \in \sigma(T) : |\lambda| \geq \frac{1}{n}\}$, $n \in \mathbb{N}$, is. This also implies that the only possible accumulation point for the spectrum in zero. To prove the claim, take a sequence

$$(\lambda_n)_n \text{ in } U_\varepsilon, \quad \text{such that} \quad \lambda_n \neq \lambda_m, \quad \forall n \neq m.$$

Consider an associated sequence of eigenvectors $(e_n)_n$, such that $Te_n = \lambda_n e_n$, for every $n \in \mathbb{N}$, $\|e_n\| = 1$. Then, the set $\{e_n, n \in \mathbb{N}\}$ is composed of linearly independent vectors. Such a family of eigenvectors exists since eigenvectors corresponding to distinct eigenvalues are linearly independent. If this family of vectors does not exist, then U_ε would be finite.

Defining now, for every $n \in \mathbb{N}$ the subspaces $E_n := \mathrm{span}(e_1, \ldots, e_n)$, we have that $E_n \subsetneq E_{n+1}$ and $T(E_n) \subseteq E_n$ from the definition of E_n. Thus, for every $n \in \mathbb{N}$ we can find $u_n \in E_{n+1}$ such that $\|u_n\| = 1$ and $\mathrm{dist}(u_n, E_n) \geq \frac{1}{2}$. Let

$$v_n := \frac{u_n}{\lambda_{n+1}}, \quad \text{thus in particular} \quad \|v_n\| \leq \frac{1}{\varepsilon}, \quad \forall n \in \mathbb{N}.$$

We now prove that, for any $n, m \in \mathbb{N}$, with $m < n$ we have

$$\|Tv_n - Tv_m\| \geq \frac{1}{2}. \tag{9.37}$$

This gives a contradiction since $\|v_n\| \leq \frac{1}{\varepsilon}$ for any $n \in \mathbb{N}$ and T is compact. To show (9.37), take any $n, m \in \mathbb{N}$ with $m < n$ and write

$$T v_n - T v_m = u_n - w_{n,m}, \quad \text{where} \quad w_{n,m} := u_n - T v_n + T v_m.$$

We conclude by showing that $w_{n,m} \in E_n$: from the choice of u_n this gives (9.37) concluding the proof. To this aim, first remark that $T v_m \in T(E_{m+1}) \subseteq E_{m+1} \subseteq E_n$, since $m < n$. Next, one can check that $u_n - T v_n = u_n - \frac{T u_n}{\lambda_{n+1}} \in E_n$, since $u_n \in E_{n+1}$ and $T u_n = \lambda_{n+1} u_n$ modulo E_n, showing that $w_{n,m} \in E_n$ and concluding the proof.

$\square$

We now conclude this section studying compact selfadjoint operators on a Hilbert space $\mathcal{H}$. As preliminary result we report the following lemma, which directly follows from previous considerations.

Lemma 9.27 *Let $\mathcal{H}$ be a Hilbert space and take $T \in \mathcal{L}(\mathcal{H})$ compact and selfadjoint. Then T has at leas a non-zero eigenvalue and $\|T\| = \max\{|\lambda|, \lambda \in \sigma_p(T)\}$. In particular, if T is positive, then $\|T\| \in \sigma_p(T)$.*

We also need the following result.

Proposition 9.28 *Let $\mathcal{H}$ be a Hilbert space and let F be a closed subspace. Given any $x \in \mathcal{H}$ let Px be the projection of x onto F. Then $P \in \mathcal{L}(\mathcal{H})$; moreover, $\|P\| \leq 1$ and $P^2 = P$.*

Proof Take any $x \in \mathcal{H}$. Using the definition of projection, Px is such that $\|x - Px\| = \mathrm{dist}(x, F)$ and $Px \in F$ and we also know that we can decompose in a unique way $x = x_1 + x_2$, with $x_1 \in F$ and $x_2 \in F^\perp$. Thus $Px = x_1$. From this we immediately check that $P(x + y) = Px + Py$ and analogously that $P(\alpha x) = \alpha Px$. Thus P is linear. Moreover, $\|Px\| = \|x_1\| \leq \|x\|$ proving continuity of P. Finally $P(Px) = P(x_1) = x_1 = P(x)$, thus $P^2 = P$ concluding the proof. $\square$

Theorem 9.29 (Spectral Decomposition) *Let $\mathcal{H}$ be a Hilbert space of infinite dimension and let $T \in \mathcal{L}(\mathcal{H})$, $T \neq 0$, be a compact and selfadjoint operator such that $Ran(T)$ is infinite dimensional. Let $\Lambda := \sigma_p(T)$ and $\Lambda^* := \sigma_p(T) \setminus \{0\}$. For any $\lambda \in \Lambda$ let E_λ be the eigenspace associated with λ and let P_λ be the projection operator on E_λ. Then the following hold:*

1. *$\Lambda \subset \mathbb{R}$ is a compact, infinite and countable set;*
2. *for every $\lambda \in \Lambda^*$, the space E_λ is finite dimensional;*
3. *for any $\lambda, \mu \in \Lambda$ with $\lambda \neq \mu$, then $E_\lambda \perp E_\mu$;*
4. *$T = \sum_{\lambda \in \Lambda^*} \lambda P_\lambda$, and in particular this series converges in $\mathcal{L}(\mathcal{H})$;*
5. *for every $x \in \mathcal{H}$*

$$Tx = \sum_{\lambda \in \Lambda^*} \lambda P_\lambda x, \quad \text{and} \quad \overline{Ran(T)} = \overline{\oplus_{\lambda \in \Lambda^*} E_\lambda}.$$

Putting together the results previously proved in this section, we only need to show that Λ^* is infinite and that item 4 holds. Nevertheless, we decided to state the result as above, since it collects the main consequences of the spectral properties of compact selfadjoint operators.

Proof We first show that Λ^* is an infinite set. To this aim, suppose by contradiction that $\Lambda^* := \{\lambda_1, \ldots, \lambda_k\}$ for some $\lambda_i \in \mathbb{R}$, $i = 1, \ldots, k$, $k \in \mathbb{N}$. First define

$$G := \oplus_{j=1}^{k} E_{\lambda_j}, \quad \text{and} \quad F := G^{\perp}.$$

Since G is finite dimensional, and thus in particular closed, we have $\mathcal{H} = G \oplus F$. Moreover, by construction, $T(G) \subset G$. Take now any $u \in G$ and $v \in F$, and using selfadjointness of T we see that

$$0 = (Tu, v) = (u, Tv) = 0, \quad \Longrightarrow \quad Tv \in F \implies T(F) \subset F.$$

Thus, considering the operator $T_F := T|_F \in \mathcal{L}(F)$, we see that it is also a selfadjoint operator and moreover,

$$T(\overline{B}_1^F) = T(\overline{B}_1^{\mathcal{H}}) \cap F,$$

proving that T_F is a compact operator on $\mathcal{L}(F)$. Finally remark that $T_F \neq 0$, since if T_F were the zero operator, then

$$T(\mathcal{H}) = T(G \oplus F) \subseteq T(G) + T(F) = T(G) + T_F(F) = T(G) \subseteq G,$$

and this would imply that $\mathrm{Ran}(T)$ is finite dimensional, contradicting the initial hypothesis. Thus, recalling Lemma 9.27, we can conclude that T_F has at least a nonzero eigenvalue μ, which of course is also an eigenvalue for T. Since $\mu \neq \lambda_j$ for all $j = 1, \ldots, k$, this gives a contradiction concluding the proof of this first point.

Finally, we show item 4. To this aim, let $A \subset \Lambda^*$ be finite and arbitrary and let

$$G_A := \oplus_{\lambda \in A} E_\lambda, \quad F_A := G_A^{\perp}.$$

Reasoning as in the previous part of the proof, we can write $\mathcal{H} = G_A \oplus F_A$ and define the compact and selfadjoint operator $T_{F_A} := T|_{F_A} \in \mathcal{L}(F_A)$. Using now Lemma 9.27, we know that

$$\left\| T_{F_A} \right\| = \max\{|\lambda| : \lambda \in \sigma(T_{F_A})\} = \max_{\Lambda^* \backslash A} |\lambda|$$

where the last equality holds since $\sigma_p(T_{F_A}) = \Lambda^* \setminus A$. Finally, considering the operator $\sum_{\lambda \in A} P_\lambda \in \mathcal{L}(\mathcal{H})$, we can write, for every $x \in \mathcal{H}$,

$$x = \underbrace{\left(x - \sum_{\lambda \in A} P_\lambda x \right)}_{\in F_A} + \underbrace{\sum_{\lambda \in A} P_\lambda x}_{\in G_A}$$

and applying the Pythagorean Theorem 5.3, we get

$$\|x\|^2 = \left\| x - \sum_{\lambda \in A} P_\lambda x \right\|^2 + \left\| \sum_{\lambda \in A} P_\lambda x \right\|^2 .$$

Thus,

$$\left\| T\left(x - \sum_{\lambda \in A} P_\lambda x \right) \right\| = \left\| T_{F_A}\left(x - \sum_{\lambda \in A} P_\lambda x \right) \right\| \le \|T_{F_A}\| \left\| x - \sum_{\lambda \in A} P_\lambda x \right\| \le \left(\max_{\Lambda^* \backslash A} |\lambda| \right) \|x\| .$$
$$\tag{9.38}$$

Finally, since $T P_\lambda = \lambda P_\lambda$ we get

$$\left\| T - \sum_{\lambda \in A} \lambda P_\lambda \right\| \le \max_{\Lambda^* \backslash A} |\lambda|.$$

Thus, for every $\varepsilon > 0$, if $A := \{\lambda \in \Lambda^* : |\lambda| \ge \varepsilon\}$, which is finite thanks to Proposition 9.26, we obtain

$$\left\| T - \sum_{\lambda \in A} \lambda P_\lambda \right\| \le \max_{\Lambda^* \backslash A} |\lambda| \le \varepsilon,$$

concluding the proof. $\qquad\square$

Corollary 9.30 *Let P_0 be the projection on $E_0 := \ker(T)$. Then*

$$\mathcal{H} = \overline{\oplus_{\lambda \in \Lambda} E_\lambda},$$

or equivalently, for every $x \in \mathcal{H}$ we can write $x = \sum_{\lambda \in \Lambda} P_\lambda x$.

The proof of this corollary consists in noticing that $\mathcal{H} = E_0 \oplus \overline{\mathrm{Ran}(T)}$, but from Theorem 9.29 we have $\overline{\mathrm{Ran}(T)} = \overline{\oplus_{\lambda \in \Lambda^*} E_\lambda}$.

Corollary 9.31 *Let $\mathcal{H}$ be a separable Hilbert space and let T be a compact and selfadjoint operator in $\mathcal{L}(\mathcal{H})$. Then there exists a countable Hilbert basis $\mathcal{E}$ made of eigenvectors of T. In particular, choosing $\mathcal{E} := \{e_n, n \in \mathbb{N}\}$ where for every $n \in \mathbb{N}$ we take e_n normalized eigenvector associated to the eigenvalue λ_n, and the eigenvalue is intended to be repeated as many times as its multiplicity (which is finite if $\lambda_n \neq 0$ and can be infinite if $\lambda_n = 0$)), then*

$$x = \sum_{n=1}^{\infty} (x, e_n) e_n, \quad and \quad Tx = \sum_{n=1}^{\infty} (x, e_n) \lambda_n e_n, \quad \forall x \in \mathcal{H}. \tag{9.39}$$

Proof Denote by $E_0 := \ker(T) = \mathrm{Ran}(T)^\perp$, which is a separable Hilbert space, and thus admits $\mathcal{E}_0$ countable Hilbert basis (see 5.20). Next consider $\mathcal{E}_\lambda$ finite Hilbert basis of E_λ, for $\lambda \in \Lambda^*$, and pose $\mathcal{E} := \cup_{\lambda \in \Lambda} \mathcal{E}_\lambda$. The thesis follows from the identity $\mathcal{H} = \overline{\oplus_{\lambda \in \Lambda} E_\lambda}$ and from orthonormality of the eigenvectors (see Corollary 9.30 and Theorem 9.29). $\qquad\square$

Proposition 9.32 (Fredholm alternative) *Let $\mathcal{H}$ be a Hilbert space and let $T \in \mathcal{L}(\mathcal{H})$ be a compact and selfadjoint operator. Let $\mu \in \mathbb{C} \setminus \{0\}$ and, for every $y \in \mathcal{H}$ consider the equation*

$$(\mu - T)x = y.$$

Then one among the following two options hold:

1. *$\mu \in \rho(T)$, (or equivalently $\mu \notin \sigma_p(T)$ from Proposition 9.26), and then there exists a unique solution given by $x = (\mu - T)^{-1}y$;*
2. *$\mu \in \sigma_p(T)$: the equation admits solution if and only if $y \in \ker(\mu - T)^\perp$ and moreover, the solutions are infinitely many and can be written as $x = \tilde{y} + z$, where $\tilde{y} \in (\ker(\mu - T))^\perp$ and $z \in \ker(\mu - T)$ is arbitrary.*

Proof The proof is a direct consequence of Theorem 9.29. The first item is immediate, whilst the second one exploits the equalities

$$\mathrm{Ran}(\mu - T) = (\ker(\mu - T))^\perp \quad \text{and} \quad \mathcal{H} = \ker(\mu - T)^\perp \oplus \ker(\mu - T),$$

which follow from the fact that $\mu I - T$ is selfadjoint and has closed range. $\qquad\square$

Appendix
Proposed Problems

1. Let X be a separable Banach space and Y a subspace of X. Show that Y, endowed with the induced norm, is separable.
2. Let X be a Banach space and Y a finite-dimensional subspace of X. Show that Y is closed.
3. Let (M, d) be a compact metric space. Show that M is complete and separable.
4. Let (M, d) be a complete metric space and $\{A_n, \ n \in \mathbb{N}\}$ a countable family of open and dense subsets of M. Show that the set $A := \bigcap_{n \in \mathbb{N}} A_n$ is dense in M.
5. Let $\mathcal{H}$ be a real Hilbert space and $a \in \mathcal{H}$ a nonzero vector. Show that for every $x \in \mathcal{H}$, we have
$$\mathrm{dist}\left(x, \{a\}^{\perp}\right) = \frac{|(x, a)|}{\|a\|}.$$

6. Consider the space ℓ^{∞} with its usual norm $\|\cdot\|_{\infty}$ and the sets $c_0 := \{(a_n)_n \in \ell^{\infty} : a_n \to 0\}$ and $c := \{(a_n)_n \in \ell^{\infty} : a_n \to a \in \mathbb{R}\}$. Show that c_0 and c are closed separable subspaces of ℓ^{∞}.
7. Consider the Hilbert space ℓ^2 and a real sequence $(a_n)_n$ such that $a_n > 0$ for every $n \in \mathbb{N}$ and $a_n \to \infty$. Show that the set
$$A := \left\{ u \in \ell^2 : \sum_{n \in \mathbb{N}} a_n \, |u_n|^2 \le 1 \right\}$$

is a precompact subset of ℓ^2.
8. Let $\mathcal{H}$ be a Hilbert space and C_1, C_2 two nonempty, closed and convex subsets such that $C_1 \subset C_2$. Given $x \in \mathcal{H}$, call $P_{C_i} x$ the projection of x on C_i and $\mathrm{d}(x, C_i)$ the distance of x from C_i $(i = 1, 2)$. Show that
$$\left\| P_{C_1} x - P_{C_2} x \right\|^2 \le 2 \left(\mathrm{d}(x, C_1)^2 - \mathrm{d}(x, C_2)^2 \right), \quad \forall x \in \mathcal{H}.$$

9. Let $\mathcal{H}$ be a complex Hilbert space and $T \in \mathcal{L}(\mathcal{H})$ an operator such that $\|T\| \le 1$. Show that:

S. Zagatti, *Functional Analysis*, SISSA Springer Series 7,
https://doi.org/10.1007/978-3-032-24475-8

(a) $Tx = x$ if and only if $(Tx, x) = \|x\|^2$;

(b) $\ker(I - T) = \ker(I - T^*)$.

10. Find a Banach space X and a subset $S \subseteq X$ such that S is strongly closed but not weakly closed.

11. Find a Banach space X, a bounded closed subset $S \subseteq X$ and a continuous function $f : S \to \mathbb{R}$ such that

$$\sup_{x \in S} f(x) = \infty.$$

12. Let X be a Banach space and $K \subseteq X$ a compact subset. Show that any sequence in K which converges weakly, actually converges strongly.

13. Let (X, d) be a metric space. Given two subsets $A, B \subseteq X$, set

$$\operatorname{dist}(A, B) := \inf\{d(x, y) : x \in A, y \in B\}.$$

(a) Given $x \in X$ and positive numbers $0 < \rho < r$, show that there exists $\delta > 0$ such that

$$\operatorname{dist}\left(B(x, \rho), B(x, r)^c\right) \geq \delta.$$

(b) Given a proper, nonempty, closed subset $C \subseteq X$, show that there exists a ball $B(x, r)$ in X such that $\operatorname{dist}(B(x, r), C) > 0$.

14. Let X, Y be Banach spaces and $T \in \mathcal{L}(X, Y)$ a compact operator. Let $(x_n)_n$ be a sequence in X weakly converging to x in X. Show that the sequence $(Tx_n)_n$ converges strongly to Tx in Y.

15. Let $\alpha > 0$ and consider the sequence of functions given by

$$u_n(x) := \min\left\{1, |x|^{-\alpha}\right\} \chi_{B(0,n)}(x), \quad n \in \mathbb{N}, x \in \mathbb{R}^d.$$

Study the convergence of $(u_n)_n$ in the strong and weak (weak* if $p = \infty$) topology of $L^p\left(\mathbb{R}^d\right)$ for $p \in [1, \infty]$.

16. Let $\mathcal{H}$ be a Hilbert space, $T \in \mathcal{L}(\mathcal{H})$ and T^* the adjoint of T.

(a) Show that $\|T^*T\| = \|TT^*\| = \|T\|^2$.

(b) Show that T^*T and TT^* are selfadjoint operators.

17. Let $\mathcal{H}$ be a Hilbert space and $\{M_k, k \in \mathbb{N}\}$ a countable collection of finite dimensional subspaces of $\mathcal{H}$. Call P_k the orthogonal projector on M_k, $k \in \mathbb{N}$ and set

$$P := \sum_{k=1}^{\infty} 2^{-k} P_k$$

Show that P is a compact operator in $\mathcal{L}(\mathcal{H})$.

18. Consider the sequence of functions given by $u_n(x, y) = \left(\cos\left(\frac{x}{n}\right) + \sin\left(\frac{x}{n}\right)\right)\left(1 + e^{-ny^2}\right)$, $(x, y) \in I := [-1, 1] \times [-1, 1]$, $n \in \mathbb{N}$. Study the convergence of $(u_n)_n$ in the strong and weak topology of $L^p(I)$ (weak* if $p = \infty$).

19. Let $\mathcal{H}$ be a complex Hilbert space, $T \in \mathcal{L}(\mathcal{H})$ and $(x_n)_n$ a sequence in $\mathcal{H}$ weakly converging to $x \in \mathcal{H}$. Show that the sequence $(T x_n)_n$ converges weakly to Tx.

20. Given $x \in \mathbb{R}$, let $B(x, 1)$ be the open unit ball of center x in $\mathbb{R}$. Consider a sequence $(x_n)_n$ in $\mathbb{R}$ and define the sequence of functions $u_n := \chi_{B(x_n,1)}$, where χ denotes the characteristic function. Study the strong and weak convergence of the sequence $(u_n)_n$ in the space $L^2(\mathbb{R})$ (that is to say establish if the sequence is converging in such topologies and, in affirmative case, find the limit), in the following cases:

(a) $x_n \to 0$

(b) $|x_n| \to \infty$.

21. Let $\mathcal{H}$ be a Hilbert space endowed with the inner product $\langle \cdot, \cdot \rangle$ and D a subset of $\mathcal{H}$ such that $\mathrm{lsp}(D)$ is dense in $\mathcal{H}$. Show that, given a bounded sequence $(x_n)_n$ in $\mathcal{H}$, such that $\langle x_n, y \rangle \to \langle x_n, y \rangle$ for any $y \in D$, then $x_n \rightharpoonup x$.

22. Let $I = [0, 1] \subseteq \mathbb{R}$ and consider the Hilbert space $X = L^2(I, \mathbb{R})$. Set

$$(Tu)(x) := \int_0^x u(t)dt$$

Show that $T \in \mathcal{L}(X)$ and find the adjoint T^* of T.

23. Consider the set $E := \{e^n, n \in \mathbb{N}\}$ in ℓ^2 defined by

$$e^n(k) = \delta_{n,k}.$$

Show that E is a Hilbert basis in ℓ^2.

24. Let $\mathcal{U}$ be a bounded family in $L^1(\mathbb{R})$ and $\rho \in C_c^\infty(\mathbb{R})$. Show that the family $\{\rho * u, u \in \mathcal{U}\}$ is equicontinuous.

25. Let $\mathcal{H}$ be a Hilbert space and $T \in \mathcal{L}(\mathcal{H})$. Show that T is compact if and only if the adjoint T^* is compact.

26. Let $\mathcal{H}$ be a Hilbert space on $\mathbb{C}$, $\{e_k, k \in \mathbb{N}\}$ an orthonormal system in $\mathcal{H}$ and $(\lambda_k)_k$ an element of $\ell^1(\mathbb{C})$. Set

$$Tx := \sum_{k=1}^{\infty} \lambda_k \, (x, e_k) \, e_k$$

Show that T is a compact operator in $\mathcal{L}(\mathcal{H})$.

27. Consider the Hilbert space $E := L^2(\mathbb{R}^n, \mathbb{C})$ and let $K \in L^2(\mathbb{R}^n \times \mathbb{R}^n, \mathbb{C})$. Define

$$(T_K u)(x) := \int_{\mathbb{R}^n} K(x, y)u(y)dy$$

Show that $T_K \in \mathcal{L}(E)$ and that T_K is selfadjoint if and only if $K(x, y) = \overline{K(y, x)}$ for any pair $(x, y) \in \mathbb{R}^n \times \mathbb{R}^n$.

28. Let $\mathcal{H}$ be a Hilbert space and $(u_n)_n$ an orthonormal sequence in $\mathcal{H}$. Show that $(u_n)_n$ converges weakly to zero.

29. Let $p \in [1, \infty[$ and $f \in L^p(\mathbb{R})$. Show that for every $\delta > 0$ we have

$$m(\{x : |f(x)| > \delta\}) \le \delta^{-p} \|f\|_p^p,$$

where m denotes the Lebesgue measure in $\mathbb{R}$.

30. Let $E \subseteq \mathbb{R}$ be a measurable set with finite measure, $p \in]1, \infty]$, $(u_n)_n$ a sequence in $L^p(E)$ and $u \in L^p(E)$ such that $u_n \rightharpoonup u$ for $p < \infty$ or $\overset{*}{\rightharpoonup}$ if $p = \infty$. Prove that the sequence $(u_n)_n$ is equi-integrable.

31. Let $\mathcal{H}$ be a real Hilbert space, $M \subseteq \mathcal{H}$ a closed subspace and P the orthogonal projector on M. Show that P is selfadjoint.

32. Consider the space $X = C_0\left(\mathbb{R}^d\right) := \overline{C_c\left(\mathbb{R}^d\right)}^{\|\cdot\|_\infty}$. Given $S \in X'$ define

$$U := \left\{ O \subseteq \mathbb{R}^d \text{ open}: \langle S, u\rangle_{X',X} = 0 \forall u \in X \text{ with } \operatorname{supp} u \subseteq O \right\}.$$

Then introduce

$$N := \bigcup_{O \in U} O \quad \text{(domain of nullity of } S\text{)}; \quad \operatorname{supp} S := \mathbb{R}^d \backslash N \quad \text{(support of } S\text{)}.$$

(i) Given $a \in \mathbb{R}^d$, set $T_a(u) = u(a)$. Show that $T_a \in X'$ and find its norm and support.

(ii) Let $(a_n)_n$ be a sequence in $\mathbb{R}^d$, consider the sequence $(S_n)_n = \left(T_{a_n}\right)_n$ and the series

$$S = \sum_{n=1}^{\infty} 3^{-n} S_n.$$

(a) Show that $S \in X'$ and find its norm and support.

(b) Show that there exists a subsequence $\left(S_{n_k}\right)_k$ weakly* converging in X'.

33. Consider the sequence of functions given by $u_n(t) = \sin(nt)$, with $n \in \mathbb{N}$ and $t \in I := [0, 2\pi]$. Study the convergence of $(u_n)_n$ in the uniform topology of $C(I)$, in the strong topology of $L^\infty(I)$ and in the weak* topology of $L^\infty(I)$ (that is to say establish if the sequence is converging in such topologies and, in affirmative case, find the limit).

34. Let $X = C_0\left(\mathbb{R}^2, \mathbb{R}\right)$, endowed with the uniform norm, and $(a_n)_n$ a sequence in $\mathbb{R}^+$. Set

$$\langle f_n, u\rangle := \int_0^{2\pi} u\left(a_n \cos\theta, a_n \sin\theta\right) d\theta, \quad \forall n \in \mathbb{N}, \forall u \in X$$

Show that $f_n \in X'$ for every $n \in \mathbb{N}$ and find its norm and support.

Suppose $a_n \to 0+$ and study the convergence of the sequence $(f_n)_n$ in the strong and weak* topology of X' (that is to say establish if the sequence converges in such topologies and, in the affirmative case, find the limit).

35. Given $\alpha \in \mathbb{R}$ and $R > 0$, consider the function u defined on $\mathbb{R}^d$ by

$$u(x) = \begin{cases} |x|^\alpha, & x \neq 0 \\ 0, & x = 0. \end{cases}$$

Establish for which $p \in [1, \infty]$ we have $u \in L^p\left(B_{\mathbb{R}^d}(0, R)\right)$.

36. Let $E \subseteq \mathbb{R}^d$ be a measurable set, $p, q \in [1, \infty[$ and $u \in L^p(E) \cap L^q(E)$. Given $\alpha \in [0, 1]$, set $\frac{1}{r} := \frac{1-\alpha}{p} + \frac{\alpha}{q}$. Show that $u \in L^r(E)$ and that

$$\|u\|_{L^r(E)} \leq \|u\|_{L^p(E)}^{1-\alpha} \cdot \|u\|_{L^q(E)}^{\alpha}.$$

37. Let $I := [0, 1]$ and consider the sequence of functions given by

$$u_n(t) = e^{-nt}, \quad t \in I, \quad n \in \mathbb{N}.$$

Study the convergence of the sequence $(u_n)_n$ in the following spaces:

 (i) $\mathcal{C}^0(I)$ endowed with the uniform topology;
 (ii) $L^1(I)$ endowed with the strong topology;
 (iii) $L^1(I)$ endowed with the weak topology;
 (iv) $L^\infty(I)$ endowed with the strong topology;
 (v) $L^\infty(I)$ endowed with the weak* topology.

38. Let $E \subseteq \mathbb{R}$ be a measurable set with fine measure and let $m \in L^\infty(E)$. Set

$$(Tu)(x) := m(x) \cdot u(x) \text{ for a.e. } x \in E.$$

Given $p, q \in [1, \infty[$, with $p \geq q$, show that $T \in \mathcal{L}\left(L^p(E), L^q(E)\right)$ and provide an estimate of its norm.

39. Let $X = \mathcal{C}_c(\mathbb{R})$ and $T : X \to X$ be a linear map such that

$$\|Tu\|_{L^1} \leq \|u\|_{L^1}; \quad \|Tu\|_{L^2} \leq \|u\|_{L^1} \quad \forall u \in X.$$

Given $r \in [1, 2]$, show that there exists $\widetilde{T} \in \mathcal{L}\left(L^1, L^r\right)$ such that $\|\widetilde{T}\|_{\mathcal{L}(L^1, L^r)} \leq 1$ and $\widetilde{T}\big|_X = T$.

40. Consider the sequence of functions given by

$$u_n(x, y) = \cos(nx)e^{-ny}, \quad (x, y) \in I := [0, 2\pi] \times [0, 2\pi], \quad n \in \mathbb{N}.$$

 (a) Study the equicontinuity of $(u_n)_n$ on I.
 (b) Study the convergence of $(u_n)_n$ in the uniform topology of $\mathcal{C}(I)$, in the strong topology of $L^\infty(I)$ and in the weak* topology of $L^\infty(I)$.

41. Let $X = C_0 \left(\mathbb{R}^2, \mathbb{R} \right)$ endowed with the uniform topology and consider the family of subsets of $\mathbb{R}^2$ given by

$$A_\alpha := \left\{ (x, y) \in \mathbb{R}^2 : y > \alpha|x|, \, x^2 + y^2 < \alpha^{-2} \right\}, \quad \alpha > 0.$$

Set $T_\alpha u := \int_{A_\alpha} u(x, y)\,dxdy, \quad \alpha > 0.$

(a) Show that $T_\alpha \in X'$ for every $\alpha > 0$ and find its norm and support.
(b) Study the convergence of the family $(T_\alpha)_{\alpha > 0}$ in the strong and weak* topology of X' when $\alpha \to 0+$ and when $\alpha \to \infty$.

42. Let $I = [0, 1]$, $X = C(I, \mathbb{R})$ and $Y = L^2(I)$. Set

$$(Tu)(x) := \int_{x^2}^{x} u(t)\,dt.$$

(a) Show that $T \in \mathcal{L}(X)$ and establish if $T\left(B_1^X\right)$ is relatively compact in X.
(b) Show that $T \in \mathcal{L}(Y)$ and establish if $T\left(B_1^Y\right)$ is relatively compact in Y.

43. Consider the sequence of functions given by

$$u_n(x, y) = \sin\left(\frac{n^2 x}{n+1} \right) e^{y/n}, \quad (x, y) \in I := [0, 2\pi] \times [0, 2\pi], \quad n \in \mathbb{N}.$$

(a) Study the equicontinuity of $(u_n)_n$ on I.
(b) Study the convergence of $(u_n)_n$ in the uniform topology of $C(I)$; in the strong and in the weak* topology of $L^\infty(I)$.

44. Let $X = C_0 \left(\mathbb{R}^2, \mathbb{R} \right)$ endowed with the uniform norm and consider the family of subsets of $\mathbb{R}^2$ given by

$$A_\alpha := \left\{ (x, y) \in \mathbb{R}^2 : x > 0, \, y > \alpha|x|, \, x^2 + y^2 < \alpha^2 \right\}, \quad \alpha > 0.$$

Set
$$T_\alpha u := \frac{1}{\alpha^2} \int_{A_\alpha} u(x, y)\,dxdy, \quad \alpha > 0.$$

(a) Show that $T_\alpha \in X'$ for any $\alpha > 0$ and find its norm and support.
(b) Establish if the family $(T_\alpha)_{\alpha > 0}$ converges in the strong and weak* topology of X' when $\alpha \to 0+$ and, in affirmative case, determine the limit T_0.
(c) Find norm and support of T_0.

45. Let $I = [0, 1]$, $X = C(I, \mathbb{R})$ and $\alpha(x) := \min\{1, 2x\}$. Set

$$(Tu)(x) := \int_0^{\alpha(x)} |u(t)|^2\,dt.$$

Establish if $T \in \mathcal{L}(X)$ and if $T\left(B_1^X\right)$ is relatively compact in X.

46. Consider the following family of Cauchy problems:

$$\begin{cases} y' = \frac{1}{1+ty} & t > 0 \\ y(0) = 1 + \frac{1}{n} & n \in \mathbb{N}. \end{cases}$$

(a) Show that for every $n \in \mathbb{N}$ there exists a solution $y_n(\cdot)$ defined on the whole $\mathbb{R}^+$.
(b) Show that the sequence $(y_n)_n$ admits a subsequence uniformly converging on each compact subinterval of $\mathbb{R}^+$.

47. Consider the sequence of functions given by

$$u_n(x, y) = \sin\left(\frac{nx}{n+1}\right)\left(1 + e^{-n|y|}\right), \quad (x, y) \in I := [-1, 1] \times [-1, 1], \quad n \in \mathbb{N}.$$

(a) Study the equicontinuity of $(u_n)_n$ on I.
(b) Study the convergence of $(u_n)_n$ in the uniform topology of $C(I)$, in the strong topology of $L^\infty(I)$ and in the weak* topology of $L^\infty(I)$.

48. Let $I = [0, 1]$, $X = C^0(I)$ and $m \in X$. Set

$$(T_m u)(x) := m(x)u(x), \quad u \in X, x \in I$$

Show that $T_m \in \mathcal{L}(X)$ and that it is compact if and only if $m(x) = 0$ for every $x \in I$.

49. Let $\overline{B}$ the closed unit ball in $\mathbb{R}$, endowed with the euclidean norm $\|\cdot\|$. Define

$$u_n(x) := |\sin(\|x\|)|^{\frac{1}{n}} \quad n \in \mathbb{N}.$$

Study the equicontinuity of the family $\{u_n, n \in \mathbb{N}\}$ on $\overline{B}$.

50. Consider the sequence of functions given by

$$u_n(x, y) = \frac{e^{-\frac{ny}{n+1}}}{\left(1 + e^{-nx^2}\right)}, \quad (x, y) \in I := [-1, 1] \times [-1, 1], \quad n \in \mathbb{N}.$$

Study the convergence of $(u_n)_n$ in the uniform topology of $C(I)$, in the strong topology and in the weak* topology of $L^\infty(I)$.

51. Let $\varphi \in C_c(\mathbb{R})$ and $(a_n)_n$ a sequence in $\mathbb{R}$. Define

$$u_n(x) := \varphi(x - a_n), \quad x \in \mathbb{R}, \quad n \in \mathbb{N}.$$

(a) Show that $u_n \in L^p(\mathbb{R})$ for every $p \in [1, \infty]$.

(b) Study the relative compactness of the sequence $(u_n)_n$ in the strong and in the weak topology of L^p (weak* if $p = \infty$). That is to say: establish if and for which $p \in [1, \infty]$ there exists a converging subsequence in such topologies.

52. Let $I = [0, 1] \subset \mathbb{R}$ and $B := \left\{ u \in C^1(I) : \|u'\|_{L^2(I)} \leq 1 \right\}$.

(a) Show that B is an equicontinuous family.
(b) Given a sequence $(u_n)_n$ in $\{ u \in B : u(0) = 0, u(1) = 1 \}$, show that there exist $u \in C^0(I)$ and a subsequence $(u_{n_k})_k$ which converges uniformly to u.
(c) Show by a counterexample that property b) does not hold in B.

53. Let $\overline{B}$ the closed unit ball in $\mathbb{R}^d$, endowed with the euclidean norm $\|\cdot\|$. Set

$$u_n(x) := e^{-n\|x\|} \quad n \in \mathbb{N}.$$

Study the equicontinuity of the family $\{u_n, n \in \mathbb{N}\}$ on $\overline{B}$.

54. Consider the sequence of functions given by

$$u_n(x, y) = \min\left\{ n, |x|^{-\frac{1}{2}} \right\} \sin\left(\frac{ny}{n+1} \right), \quad (x, y) \in I := [-1, 1] \times [-1, 1], \quad n \in \mathbb{N}.$$

Study the convergence of $(u_n)_n$ in the strong and weak topology (weak* * if $p = \infty$) of $L^p(I)$.

55. Let $\varphi \in C_c(\mathbb{R})$ with $\operatorname{supp} \varphi \subseteq [-1, 1]$, $\varphi \geq 0$ and $\int_{\mathbb{R}} \varphi \, dt = 1$. Consider the Dirac sequence given by

$$\rho_n(t) := n\varphi(nt) \quad \forall t \in \mathbb{R} \quad \forall n \in \mathbb{N}.$$

and let $(a_n)_n$ be a sequence in $\mathbb{R}$. Set

$$u_n(t) := \rho_n(x - a_n) \quad \forall t \in \mathbb{R} \quad \forall n \in \mathbb{N}.$$

(a) Show that $u_n \in L^p(\mathbb{R})$ for every $p \in [1, \infty]$ and for every $n \in \mathbb{N}$.
(b) Considering the cases $a_n = n$ and $a_n = n^{-2}$, study the convergence of the sequence $(u_n)_n$ in the strong and weak topology of L^p (weak* if $p = \infty$).

56. Let $I = [0, 1] \subset \mathbb{R}$, $X = C^0(I)$ and $Y = L^1(I)$. Set

$$Tu(x) := \int_0^x xyu(y)dy$$

(a) Show that $T \in \mathcal{L}(X)$ and $T \in \mathcal{L}(Y)$.
(b) Establish if T is compact in $\mathcal{L}(X)$ and in $\mathcal{L}(Y)$, explaining the reasons.

57. Let $(\rho_n)_n$ be a regularizing sequence in $\mathbb{R}$. Study the equiintegrability of the following families:

(a) $f_n = \rho_n, \quad n \in \mathbb{N}$

(b) $g_n = \rho_n'$, $\quad n \in \mathbb{N}$;

(c) $h_n := \rho_1 \star \rho_n$, $\quad n \in \mathbb{N}$.

58. Let $\alpha > 0$ and consider the sequence of functions given by

$$u_n(x) := \min\left\{1, |x|^{-\alpha}\right\} \chi_{B(0,n)}(x), \quad n \in \mathbb{N}, x \in \mathbb{R}^d.$$

Study the strong and weak (weak* if $p = \infty$) convergence of $(u_n)_n$ in the spaces $L^p\left(\mathbb{R}^d\right)$ for $p \in [1, \infty]$.

59. Let $Q := [-1, 1]^3 \subseteq \mathbb{R}^3$ and set

$$f(x_1, x_2, x_3) = \begin{cases} \left(x_1 x_2^2 x_3^3\right)^{-1}, & x_1 x_2 x_3 \neq 0 \\ 0, & x_1 x_2 x_3 = 0. \end{cases}$$

- Establish for which $p \in [1, \infty]$ we have $f \in L^p\left(\mathbb{R}^3\right)$;
- establish for which $p \in [1, \infty]$ we have $f \in L^p(Q)$;
- establish for which $p \in [1, \infty]$ we have $f \in L^p\left(\mathbb{R}^3 \backslash Q\right)$.

60. Let $E \subseteq \mathbb{R}$ be a measurable set, $p_i \in [1, \infty]$, $f_i \in L^{p_i}(E)$ for $i = 1, \ldots, n$, and $r \in [1, \infty]$ given by

$$\frac{1}{r} := \sum_{i=1}^{n} \frac{1}{p_i}.$$

Show that

$$\prod_{i=1}^{n} f_i \in L^r(E)$$

and that the following inequality holds:

$$\left\| \prod_{i=1}^{n} f_i \right\|_{L^r(E)} \leq \prod_{i=1}^{n} \| f_i \|_{L^{p_i}(E)}.$$

61. Let $(\rho_n)_n$ be a regularizing family in $\mathbb{R}$ and $f \in C^0(\mathbb{R})$. Set

$$f_n(x) := (\rho_n * f)(x), \quad x \in \mathbb{R}.$$

Show that f_n is well-defined and that the sequence $(f_n)_n$ converges uniformly to f on any compact subset $K \subseteq \mathbb{R}$.

62. Let $I = [-1, 1] \subseteq \mathbb{R}$ and $(u_n)_n$ a sequence in $C^2(\mathbb{R})$ such that

(a) u_n is convex on $\mathbb{R}$ for every $n \in \mathbb{N}$;

(b) There exists $K \geq 0$ such that $|u_n(0)| + |u_n'(t)| \leq K$ for every $t \in I$ and for every $n \in \mathbb{N}$.

(1) Show that the sequence $(u_n')_n$ is relatively compact in $L^1(I)$.

(2) Show that there exists a subsequence $(u_{n_k})_k$ and a map $u \in C^0(I)$ such that $(u_{n_k})_k$ converges uniformly to u on I.

63. Let $I = [0, 1] \subseteq \mathbb{R}$ and $\{e_n, n \in \mathbb{N}\}$ a Hilbert basis $L^2(I)$. Set

$$(e_m \otimes e_n)(x, y) := e_m(x)e_n(y); \quad m, n \in \mathbb{N}, (x, y) \in I \times I.$$

Show that the family $\{e_m \otimes e_n; m, n \in \mathbb{N}\}$ is a Hilbert basis in $L^2(I \times I)$.

64. Let $I := [-1, 1] \subseteq \mathbb{R}$ and consider the sequence of functions given by:

$$u_n(t) = e^{-n} \cdot e^{nt^2}; \quad t \in I, \quad n \in \mathbb{N}.$$

Study the convergence of the sequence $(u_n)_n$ in the following spaces:

(i) $C^0(I)$ with uniform topology;

(ii) $L^1(I)$ with strong topology;

(iii) $L^1(I)$ with weak topology;

(iv) $L^\infty(I)$ with strong topology;

(v) $L^\infty(I)$ with weak* topology.

65. For every $n \in \mathbb{N}$ set

$$f_n(x) = \sin\left(\frac{x}{n}\right); \quad g_n(x) = \sin\left(n^2 x\right); \quad h_n(x) = \sin\left(\frac{nx}{n+1}\right); \quad x \in [0, 2\pi].$$

Study the equicontinuity of the sequences $\{f_n, n \in \mathbb{N}\}, \{g_n, n \in \mathbb{N}\}$ e $\{h_n, n \in \mathbb{N}\}$ over $[0, 2\pi]$.

66. Let $I = [0, 1] \subset \mathbb{R}$ and, for every $n \in \mathbb{N}$, consider the subintervals of the form

$$I_n^m := \left[\frac{m}{n}, \frac{m+1}{n}\right[, \quad m = 0, 1, \ldots, n - 1.$$

Then set

$$u_n(t) := (-1)^m \text{ for } t \in I_n^m.$$

Study the strong and weak convergence of the sequence $(u_n)_n$ in $L^2(I)$.

67. Let D be te unit disk in $\mathbb{C}$. Study the equicontinuity of the following families of functions in $C(D)$:

(i) $\left\{ f_a(z) = e^{iaz}, a \in \mathbb{R} \right\}$;

(ii) $\left\{ f_a(z) = e^{i\frac{z}{a}}, a \in \mathbb{R} a \neq 0 \right\}$

(iii) $\left\{ f_a(z) = e^{iaz}, a \in \mathbb{R}, |a| > 1 \right\}$;

(iv) $\left\{ f_a(z) = e^{iaz}, a \in \mathbb{R}, |a| < 1 \right\}$.

68. Let $X = \mathcal{C}([0, 1], \mathbb{R})$ and $(a_n)_n$ a sequence in $[0, 1]$. Set

$$\langle f_n, u \rangle := u\,(a_n)\,, \forall n \in \mathbb{N}, \forall u \in X.$$

Show that $f_n \in X'$ for every $n \in \mathbb{N}$ and that there exists a subsequence $\left(f_{n_k}\right)_k$ which converges in the topology $\sigma\left(X', X\right)$.

69. Study the equicontinuity of the following families in $\mathcal{C}(I)$, $(I \subseteq \mathbb{R})$:

   (i) $\{f_a(x) = e^{ax}, a \in \mathbb{R}\}$, $I = \mathbb{R}$;
   (ii) $\{f_a(x) = a(1 - x)^2, a \in \mathbb{R}^+\}$, $I = [-1, 1]$;
   (iii) $\{f_a(x) = x^{-a}, a \in \mathbb{R}^+,\}$, $I =]1, \infty[$;
   (iv) $\{f_a(x) = x^{-a}, a \in \mathbb{R}^+,\}$, $I =]0, \infty[$.

70. Let $p \in [1, \infty[$. Consider the space $X = L^p([0, 1])$ and set

$$(Tu)(x) = \int_0^x u(t)dt.$$

   (i) Show that $T \in \mathcal{L}(X)$ and that $\|T\|_{\mathcal{L}(X)} \leq \left(p^{\frac{1}{p}}\right)^{-1}$.) Given a sequence $(u_n)_n$ in X weakly converging to u in X, show that the sequence $(Tu_n)_n$ converges strongly to Tu in X.

71. Let $C > 0$, $p \in [1, \infty[, \alpha \in]0, 1[$ and $B := \{x \in \mathbb{R}^d : \|x\| \leq 1\}$. Consider the set

$$U := \{u \in C(B) : u(0) = 0, |u(x) - u(y)| \leq C|x - y|^\alpha \; \forall x, y \in B\}.$$

   Show that U is relatively compact in $L^p(B)$.

72. Let $I := [0, 1]$ and $(u_n)_n$ a sequence in $\mathcal{C}^1([0, 1])$ such that

$$|u_n(0)| + \int_I \left|u'_n(t)\right| dt \leq 1 \quad \forall n \in \mathbb{N}.$$

   Show that there exist a subsequence $\left(u_{n_k}\right)_k$ and a map $u \in L^1(I)$ such that $u_{n_k} \to u$ strongly in $L^1(I)$.

73. Let $E \subseteq \mathbb{R}^d$ be a measurable set such that $0 < m(E) < \infty$. For every $p \in [1, \infty[$ and for every $f \in L^p(E)$ set

$$N_p[f] := \left(\frac{1}{m(E)} \int_E |f(x)|^p\right)^{\frac{1}{p}}.$$

   Show that $N_p[\cdot]$ is a norm over $L^p(E)$ and that, if $1 \leq p \leq q < \infty$, we have

$$N_p[f] \leq N_q[f] \quad \forall f \in L^q(E).$$

74. Let X be a Banach space and set $\mathcal{K}(X) := \{T \in \mathcal{L}(X) : T \text{ is compact }\}$. Show that $\mathcal{K}(X)$ is closed in $\mathcal{L}(X)$.

75. Let $X = C^0\left(\mathbb{R}^2\right)$ and $(a_n)_n$ a sequence in $\mathbb{R}^+$. For every $n \in \mathbb{N}$ and for every $u \in X$ set

$$T_n(u) = \int_{-a_n}^{+a_n} u(x, nx)dx.$$

Show that $T_n \in X'$ for every $n \in \mathbb{N}$ a find its norm and support. Study the convergence of the sequence $(T_n)_n$ in the strong and weak* topology of X' in the cases $a_n = 1 + n^2$ and $a_n = e^{-\frac{1}{n}}$.

76. Let $I = [0, 1]$ and H an equicontinuous subset of $C^0(I)$. Show that $\overline{H}$ is equicontinuous.

77. Let $I = [0, 1]$, $B_r = B(0, r)$ the ball in $\mathbb{R}^d$ of center zero and radius r, $p \in [1, \infty]$, $X_p := L^p(B_1)$ and $Y := C^0(I, \mathbb{R})$. Given $u \in X_p$ and $t \in I$, set

$$(Tu)(t) := \int_{B_t} u(y)dy.$$

Show that $T \in \mathcal{L}\left(X_p, Y\right)$ for every p and establish for which p it is compact.

78. For $(x, y) \in I := [-1, 1] \times [-1, 1]$, consider the sequence of functions given by

$$u_n(x, y) = \left(\cos\left(\frac{nx^2}{n+1}\right)\sin(nx)\right)\left(1 + e^{-ny^2}\right), \quad n \in \mathbb{N}.$$

Study the convergence of $(u_n)_n$ in the strong and weak topology (weak* if $p = \infty$) of $L^p(I)$.

79. Let $(a_n)_n$ and $(b_n)_n$ sequence in $\mathbb{R}^+$ and set $R_n := [-a_n, a_n] \times [-b_n, b_n] \subseteq \mathbb{R}^2$ and

$$u_n(x, y) := \chi_{R_n}(x, y), \quad (x, y) \in \mathbb{R}^2.$$

Study the convergence of $(u_n)_n$ in the strong and weak topology of $L^1\left(\mathbb{R}^2\right)$ and in the strong and weak* topology of $L^\infty\left(\mathbb{R}^2\right)$ in the following cases:

a. $a_n = n, b_n = n^{-1}$
b. $a_n = n, b_n = n^{-\frac{1}{2}}$;
c. $a_n = \frac{n}{n+1}, b_n = n^{-1}$
d. $a_n = \frac{n}{n+1}, b_n = \frac{n}{n+1}$.

80. Let $X = C^0\left(\mathbb{R}^2, \mathbb{R}\right)$, endowed with the uniform norm, and $(a_n)_n$, $(b_n)_n$ sequences in $\mathbb{R}^+$. Define

$$\langle f_n, u \rangle := \int_0^{2\pi} u\left(a_n \cos\theta, b_n \sin\theta\right) d\theta, \quad n \in \mathbb{N}, u \in X.$$

Show that $f_n \in X'$ for every $n \in \mathbb{N}$ and find its norm and support.

Suppose $a_n \to 1$, $b_n \to 0$ and study the convergence of the sequence $(f_n)_n$ in the strong and weak* topology of X'.

81. Let $I = [0, 1]$, $M > 0$ and $(u_n)_n$ a sequence in $\mathcal{C}^1(I)$ such that

   a. $\int_I |u_n(t)|^2 \le M \ \forall n \in \mathbb{N}$;
   b. $u'_n(t) + t \ge 0 \ \forall t \in I, \forall n \in \mathbb{N}$.

   Show that the sequence $(u_n)_n$ is relatively compact in $L^1(I)$.

82. Let $(x_n)_n$ be a sequence in a Hilbert space $\mathcal{H}$ endowed with the inner product $\langle \cdot, \cdot \rangle$. Show that, if the sequence $(\langle x_n, y \rangle)_n$ converges for every $y \in \mathcal{H}$, then the sequence $(x_n)_n$ converges weakly.

83. Let $I = [0, 1]$ and call X the Banach space $\mathcal{C}(I)$, endowed with the uniform norm. Introduce the space

$$Y := \left\{ u \in X, u \text{ differentiable on } I \text{ with } u' \in X \right\}$$

   and set
$$\|u\|_Y := \|u\|_\infty + \|u'\|_\infty, \ u \in Y.$$

   Prove that $(Y, \| \cdot \|_Y)$ is a Banach space. Let α be a nonzero element of X and set

$$(Tu)(x) := \alpha(x) u'(x) \quad u \in Y, x \in I.$$

   (i) Prove that $T \in \mathcal{L}(Y, X)$ and find its norm.
   (ii) Establish if T is compact and justify the answer.

84. Let $\mathcal{H}$ be a Hilbert space. For $T \in \mathcal{L}(\mathcal{H})$ denote by $R(T)$ and $N(T)$, respectively, the range and the kernel of T. Calling T^* the adjoint of T, prove that $N(T) = (R(T^*))^\perp$ and $\overline{(R(T))} = (N(T^*))^\perp$.

85. Let $B_r = B(0, r)$ be the ball in $\mathbb{R}^d$ of center zero and radius r and $X := C_0(\mathbb{R})$. Let m be a map in $C(\mathbb{R})$, with $m(x) \ge 0$ for every $x \in \mathbb{R}$, and, for every $t > 0$, set

$$T_t(u) := t^{-d} \int_{B_t} m(y) u(y) dy.$$

   Prove that $T_t \in X'$ for every $t > 0$ and find its norm and support. Study the convergence of T_t as $t \to 0^+$ in the strong and weak* topology of X'.

86. Let $I = [0, 1]$ and $(u_n)_n$, $(v_n)_n$ be two bounded sequences in $L^2(I)$. Assume in addition that the maps $I \ni x \mapsto u_n(x)$ and $I \ni x \mapsto v_n(x)$ are continuous and monotone non decreasing for every $n \in \mathbb{N}$; then define

$$f_n(x, y) := u_n(x) v_n(y), \quad (x, y) \in Q := I \times I.$$

   Prove that f_n lies in $L^2(Q)$ for every $n \in \mathbb{N}$ and that the sequence $(f_n)_n$ is relatively compact in $L^1(Q)$.

87. Let $I = [0, 1]$, $Q := I \times I$ and $(a_n)_n$, $(b_n)_n$ sequences in $]0, 1]$. Define the family of sets $R_n := [0, a_n] \times [0, b_n] \subseteq Q$ and set

$$u_n(x, y) := (1 + \sin(nx)) \left(1 + e^{-ny}\right) \chi_{R_n}(x, y), \quad (x, y) \in Q.$$

Study the convergence of $(u_n)_n$ in the strong and weak topology of $L^1(Q)$ and in the strong and weak* topology of $L^\infty(Q)$ in the following cases:

a. $a_n = n^{-2}$, $b_n = 1 - n^{-1}$;
b. $a_n = 1 - n^{-2}$, $b_n = 1 - n^{-1}$.

88. Let $\mathcal{H}$ be a complex Hilbert space with inner product $(\cdot, \cdot)$. Prove that we have

$$4(x, y) = \left(\|x + y\|^2 - \|x - y\|^2\right) - i \left(\|x + iy\|^2 - \|x - iy\|^2\right) \quad \forall x, y \in \mathcal{H}.$$

89. Let $I = [0, 1]$ and call X the Banach space $\mathcal{C}(I)$, endowed with the uniform norm. Let $g \in \mathcal{C}(I \times I)$ and set

$$(Tu)(x) := \int_I g(x, y)u(y)dy \quad u \in X, x \in I.$$

(i) Prove that $T \in \mathcal{L}(X)$ and estimate its norm.
(ii) Establish if T is compact and justify the answer.
(iii) Compute the norm of T in the case $g(x, y) = e^{x+y}$.

90. Let $X = C_0\left(\mathbb{R}^2\right)$ and, for every $n \in \mathbb{N}$, consider the set

$$R_n :=]-n, n[\times]-n^{-1}, n^{-1}[\subseteq \mathbb{R}^2 .$$

Given $u \in X$ and $n \in \mathbb{N}$ set

$$(T_n u)(x) = \frac{1}{n} \int_{R^n} e^{-(x^2+y^2)}u(x, y)dxdy.$$

Prove that $T_n \in X'$ for every $n \in \mathbb{N}$ and find its norm and support. Study the convergence of the sequence $(T_n)_n$ in the strong and weak* topology of X'.

91. Let $Q = [0, 1]^d \subseteq \mathbb{R}$ and consider $(u_n)_n$, $(v_n)_n$, two relatively compact sequences in $L^2(Q)$. Define

$$f_n(x) := u_n(x)v_n(x), \quad x \in Q, n \in \mathbb{N}.$$

Prove that f_n lies in $L^1(Q)$ for every $n \in \mathbb{N}$ and that the sequence $(f_n)_n$ is relatively compact in $L^1(Q)$.

92. Let $\varphi : \mathbb{R}^+ \to \mathbb{R}^+$ be map of class C^1 such that $\varphi(0) = 0$ and $1 \leq \varphi'(t) \leq 2$ for every $t > 0$. Let $I = [0, 1]$ and $(u_n)_n$ a sequence in $L^1(\mathbb{R})$.

(i) Prove that the sequence $(v_n)_n$ defined by $v_n(t) := u_n(\varphi(t))$ for $t \in I$ and $n \in \mathbb{N}$ lies in $L^1(I)$.
(ii) Assuming that $u_n \to u$ strongly in $L^1(\mathbb{R})$, study the convergence of $(v_n)_n$ in the strong and weak convergence of $L^1(I)$.
(iii) Assuming that $u_n \rightharpoonup u$ weakly in $L^1(\mathbb{R})$, study the convergence of $(v_n)_n$ in the strong and weak convergence of $L^1(I)$.

93. Let $Q = [0, 1] \times [0, 1]$ and X the Banach space $C^0(Q)$, endowed with the uniform norm. Set

$$(T_n u) := \int_0^1 n e^{-nx} u\left(x, x^2\right) dx, \quad u \in X.$$

Prove that $T_n \in X'$ for every $n \in \mathbb{N}$ and find its norm and support. Study the convergence of $(T_n)_n$ in the strong and weak* topology of X'.

94. Let $\mathcal{H}$ be a Hilbert space, $T \in \mathcal{L}(\mathcal{H})$ and $(T_n)_n$ a sequence in $\mathcal{L}(\mathcal{H})$.

(i) Prove that $T_n \to T$ if and only if $T_n^* \to T^*$.
(ii) Prove that the sequence $(T_n x)_n$ converges weakly to Tx for every $x \in \mathcal{H}$ if and only if the sequence $\left(T_n^* x\right)_n$ converges weakly to $T^* x$ for every $x \in \mathcal{H}$.

95. Let $I = [0, 1] \subseteq \mathbb{R}$ and $X = C^0(I)$. Given a map $m \in L^2(I)$, set

$$Tu(x) := \int_0^{x^2} m(y)u(y)dy.$$

Prove that $T \in \mathcal{L}(X)$ and establish if T is compact in $\mathcal{L}(X)$, justifying the answer.

96. Let $Q = [0, 1]^d \subseteq \mathbb{R}$. Consider two relatively compact families U and V in $C^0(Q)$ and define

$$F := \{f : f(x) = \sin(u(x) \cdot v(x)), x \in Q, u \in U, v \in V\}.$$

Prove that F is a relatively compact family in $C^0(Q)$.

97. Let $I = [0, 1] \subseteq \mathbb{R}$, $p > 1$ and $X = L^\infty(I)$. Given a map $m \in L^p(I)$, set

$$Tu(x) := \int_0^x m(y)u(y)dy.$$

Prove that $T \in \mathcal{L}(X)$ and establish if T is compact in $\mathcal{L}(X)$, justifying the answer.

98. Let X be the Banach space $C_0\left(\mathbb{R}^2\right)$, endowed with the uniform norm, and let $(g_n)_n$ be a sequence in $C_b\left(\mathbb{R}^2\right)$ such that

$$0 \leq g_n(x, y) \leq \left(1 + x^2 + y^2\right)^{-1} \quad \forall (x, y) \in \mathbb{R}^2, \forall n \in \mathbb{N}.$$

and $g_n \to g$ in $C_b(\mathbb{R}^2)$. Set

$$(T_n u) := \int_{\mathbb{R}} g_n(x, x) u(x, x) dx, \quad u \in X.$$

Prove that $T_n \in X'$ for every $n \in \mathbb{N}$ and find its norm and support. Study the convergence of $(T_n)_n$ in the strong and weak* topology of X'.

99. Let $f \in L^2(\mathbb{R})$ and set

$$(Tu)(x) := \int_{\mathbb{R}} f(x - y) u(y) dy.$$

Establish for which indices $p, q \in [1, \infty]$ we have $T \in \mathcal{L}(L^p(\mathbb{R}), L^q(\mathbb{R}))$.

100. Let $I = [0, 1] \subseteq \mathbb{R}$, $X = C^0(I)$ and $Y = L^1(I)$. Set

$$Tu(x) := \int_0^x xy u(y) dy.$$

a) Prove that $T \in \mathcal{L}(X)$ and $T \in \mathcal{L}(Y)$.
b) Establish if T is compact in $\mathcal{L}(X)$ and in $\mathcal{L}(Y)$, justifying the answer.

Index

© The Editor(s) (if applicable) and The Author(s), under exclusive license to Springer Nature Switzerland AG 2026
S. Zagatti, *Functional Analysis*, SISSA Springer Series 7,
https://doi.org/10.1007/978-3-032-24475-8